To all my dear readers, I give the following medical disclaimer:

This book does not constitute medical advice. The remedies described here should only be taken under the Guidance of a Naturopathic Doctor or an experienced Homeopathic practitioner. Should the reader decide to take remedies on their own, the author holds no responsibility or liability for the consequences.

© 2017 Dr. Paul Anderson Theriault, BSc, ND, Naturopathic Doctor
No extracts or quotations of this work may be made without attribution to the author. Extended quotations require written permission from the author. Please write to drpaultheriault.nd@gmail.com for permission.
Published By Lulu.com

Table of Animals-Porifera, Cnidaria, Ctenophora
© Dr. Paul Theriault 2017

This book is dedicated to everyone who moves their profession forward through creating new knowledge, advancing techniques, or improving the process of human knowledge.

To Jan Scholten, Michal Yakir, Rajan Sankaran, Divya Chhabra, Massimo Mangialavori, Louis Klien, Tinus Smits and all others who have advanced homeopathy through exploring the different groups of remedies and their common characteristics. In The Cnidarian remedies, I must especially thank Jo Evans, the first person to write on the Cnidarians.

To all those who have advanced homeopathy through conducting provings and increasing our repertoire of remedies.

I would also particularly like to thank Richard Beamish for his assistance with editing this manuscript, and to Dr Ghanshyam Kalathia for his contributions to this book from a sensation perspective. And the dearest thank you to Joey Wargachuk for your work on design for this project.

Table of Animals-Porifera, Cnidaria, Ctenophora
© Dr. Paul Theriault 2017

Table of Contents

Introduction to Porifera	1
Ecological Role of Sponges	5
Unique Characteristics of Sponges	7
History of Sponge Use in Homeopathy and Medicine	7
Systematics and Chatacteristics of Porifera in Homeopathy	7
<u>Chart:</u> Themes of the Sponge Classes	9
The Sensation of Sponges- Dr Ghanshyam Kalathia	9
Proven and Unproven Remedies within Porifera	17
Proving Suggestions	18
Badiaga	19
Trituration of Badiaga: Why do I have to be Alive?	21
Euplectella aspergillum	27
Spongia toasta	30
<u>Chart:</u> Summary of Key Issues in Sponges	34
Introduction to the Cnidarians	35
Ecology of Cnidarians	40
Evolution of Cnidarians	40
Unique Characteristics of Cnidarians	41
Human Uses of Cnidarians in Food and Medicine	41
<u>Chart:</u> Themes of Cnidarian Classes	42

Systematics and Characteristics of Cnidarians in Homeopathy	42
Sensations of Cnidarians- by Dr Ghanshyam Kalathia	43
Proved and Unproved Remedies within Cnidaria	50
Proving suggestions	51
Anthopleura xanthogrammica	52
Corallium nigrum	55
Corallium rubrum	59
Corallium rubrum Trituration: Unification of the Self	60
Diploria clivosa	64
Fossil Dimorphastrea	69
Fossil Fungia Coral	71
Heteractis malu	73
Stichodactyla gigantea	75
Chironex fleckeri	77
Medusa	79
Physalia physalis	81
<u>Chart</u>: Cnidarian Remedy Summary	85
Introduction to the Ctenophores	86
Ecology of Ctenophores	89
Evolution of Ctenophores	90
Unique Characteristics of Ctenophores	91
Ctenophores in Homeopathy	91

Table of Animals-Porifera, Cnidaria, Ctenophora
© Dr. Paul Theriault 2017

Sensation of Ctenophora- by Dr Ghanshyam Kalathia	93
Suggested Provings of Ctenophores	96
Mnemiopsis macrydi	97
Mnemiopsis macrydi trituration	98
Appendix A: <u>Charts</u>	110
<u>Sponge Charts</u>: Themes, Classes, Proved and Unproved remedies	110
Summary of Key Issues of Sponges	111
<u>Ctenophora Charts</u>: Themes, Proved and Unproved Remedies	112
Cnidarian Charts: Themes, Classes, Proved and Unproved remedies	113
Summary of Key Issues of Cnidarians	114
Appendix B: <u>Table of Animals Charts</u>	115

Porifera

Porifera are the first beings we may truly call animals. Porifera are a large, ancient and diverse group of aquatic organisms ranging throughout most shallow aquatic habitats on the planet. They are commonly referred to as sponges in English.

Sponges are likely the immediate descendants of the choanoflagellates, the immediate ancestors of the animals. Some cells in fact still bear a striking resemblance to the choanoflagellates, which unfortunately remain homeopathically unproven.

Sponges are relatively simple animals. While in most animals cells lose their ability to change their external forms and functions after their differentiation from stem cells, many sponge cells (such as choanocytes or archaeocytes[1]) retain this ability (called totipotency). Sponges also lack specialized tissues and many of the features otherwise universal in animals, such as digestive organs, symmetry of any kind, and nervous systems[2].

The basic cellular structure of sponges is relatively simple. It is shown diagrammatically in Figure 1[3]:

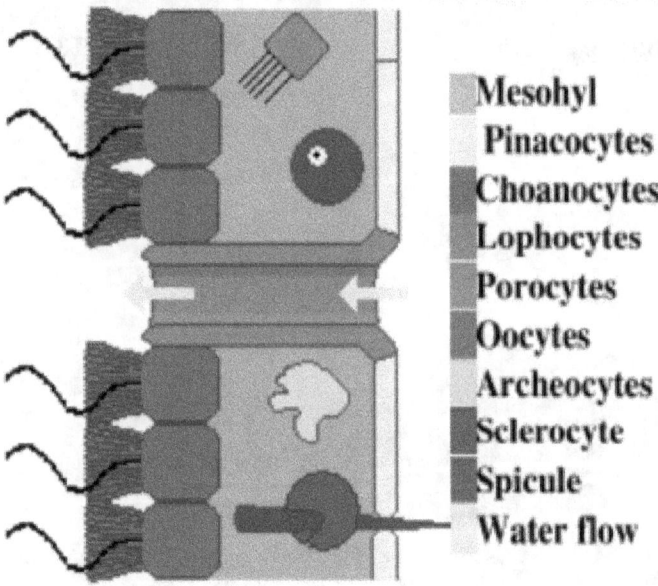

Figure 1: Sponge Cellular Structure

Sponges' cellular structure involves two layers of cells linked together by a central protein mesh called a mesophyl or mesenchyme[4]. The outer cells, or Pinacocytes, form a skin to protect the external surfaces of the sponge[5]. Inner cells called choanocytes line the inner areas of the sponge, and have flagella with which they move water through the sponge body pores[6]. These cells are also responsible for filtering food out of the water moving through the sponge, where it is digested by the totipotent archaeocytes[7]. These cells bear a striking resemblance to choanoflagellates, leading to the knowledge of the close evolutionary relationship between these two groups of organisms[8]. Other cells called myocytes and porocytes, while not muscle cells, create rudimentary contractile movement[9].

The above structure is anchored upon a hard skeleton that is secreted by sclerocytes[10]. Several different variation of skeletons are known among sponges, leading to their major classifications, discussed below.

Sponges' external morphology is also very simple, diagramed in Figure 2[11]:

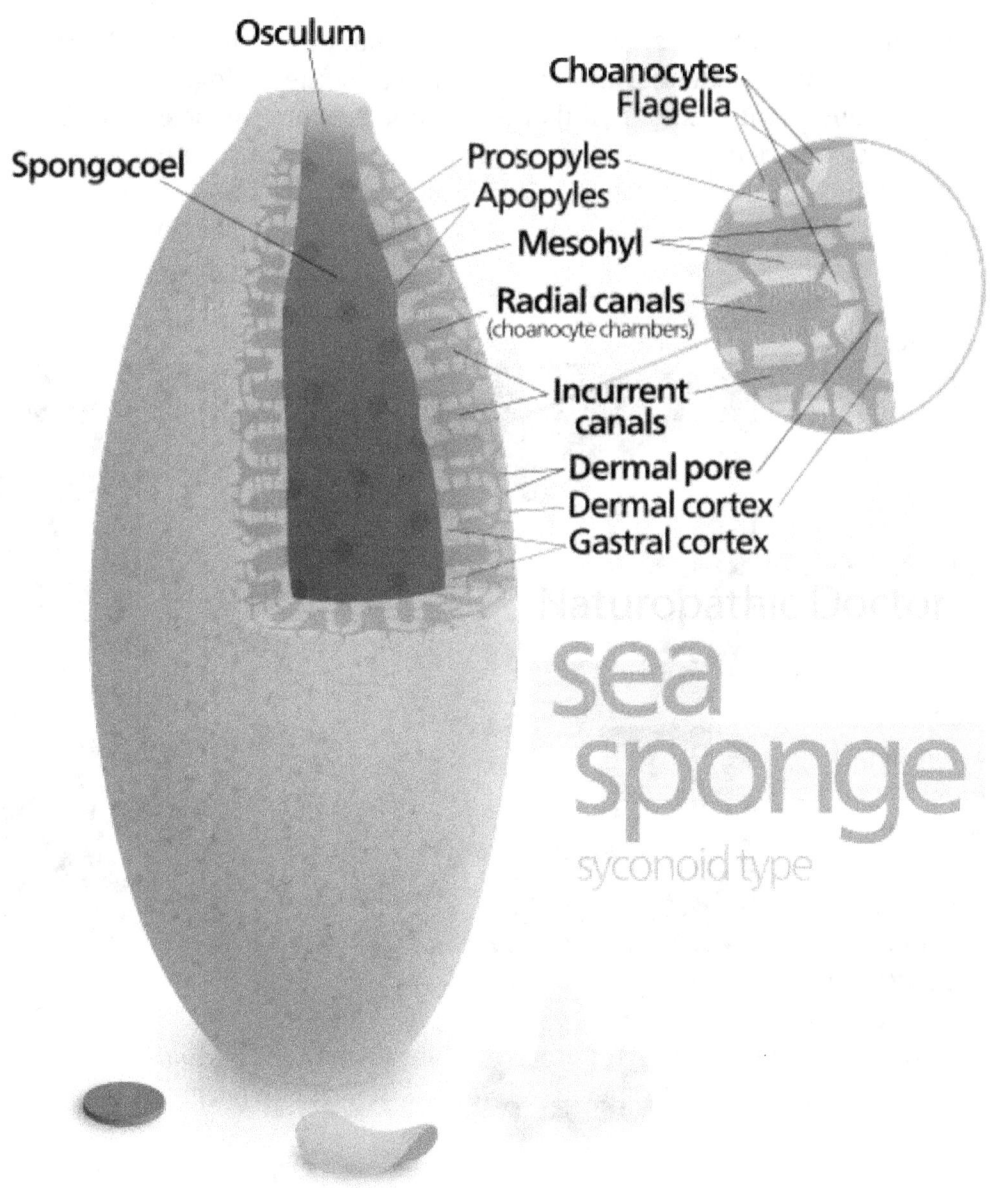

Figure 2: Sponge Morphology

Table of Animals-Porifera, Cnidaria, Ctenophora
© Dr. Paul Theriault 2017

Sponges have generally round tube-shaped bodies, which contain many small pores. Choanocytes using their flagella create a flow of water through these pores, and filter out any bacteria and small organisms that float within their reach. The filtered water collects in an internal cavity (known as a spongoceol), and is discharged out an orifice (called an osculum).

Sponges can reproduce both sexually and asexually through budding. Sexually reproducing sponges usually are hermaphroditic, and produce eggs and sperm at different times[12]. Some sponges actually capture sperm, allowing their eggs to be fertilized internally, and either release or maintain the resulting larvae[13]. Once released, larvae find a spot to settle, and mature into juvenile sponges[14].

Sponges are divided into three major classes by most modern biologists[15]. Each class has a distinct type of skeletal framework as well as some biological differences. The calcareous sponges, the hexactinellid or glass sponges, and the desmosponges[16]. Three extinct divisions are also noted; known as the archaeocyantha and stromatoporids which went extinct in the Cambrian and the Carboniferous respectively; and the sphinctozoans which went extinct at the end of the Cretaceous[17].

Figure 3: *Aphrocallistes vastus*

Arguably the earliest group of sponges is the hexactinelid or glass sponges. These sponges have skeletons composed largely of six pointed silica spicules. They are largely cup or glass shaped, and their silica skeletons are very rigid[18]. They tend to live at greater depth than other sponge classes, being found often several hundred meters below the surface. Their cells tend to form large syncytia, or large giant cells with many nuclei formed by the merging together of several small individual nucleated cells[19]. A photo of a representative species (*Aphrocallistes vastus*) is given in Figure 3 [20].

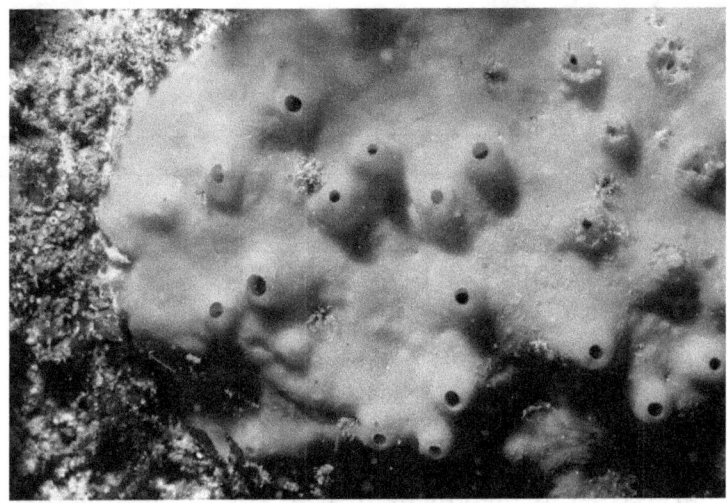

Figure 4: *Acarnus erithacus*, The Red Volcano

The second class of sponges is the desmosponges, This group is the largest in terms of species, containing approximately 90% of all sponges. This subphyla is relatively diverse, including many large sponge species. These sponges have single nucleated cells, and have skeletons made out of silica, spongin or a combination of the two[21]. They reproduce both sexually and asexually and tend to live in shallow waters. An example is shown in Figure 4 of the red volcano sponge (*Acarnus erithacus*)[22]:

The final, and most advanced, class of sponges is the calcareous sponges. Their skeletons are composed of calcium carbonate and their cells have single nuclei[23]. According to genetic analysis, these sponges are very closely related to cnidaria, and are the probable ancestors of the other animal phyla[24]. An example is given in Figure 5 of the species *Clathrina clathrus*, which has been potentised[25].

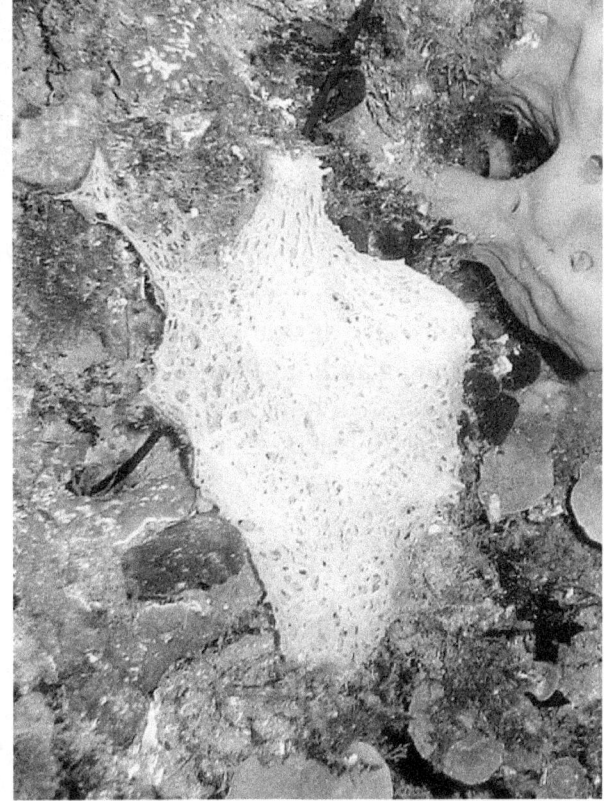

Figure 5: *Clathrina clathrus*

The extant sponges are differentiated in the chart below:

Sponge Class	Defining Characteristics
Calcareous	Skeletons made of calcium carbonate[26]. Very closely related to Cnidaria, probable ancestors of more advanced animal life.
Desmosponges	Skeletons composed of spongin (a protein) or silica. If silica skeleton is present, the structure has 1-4 rays, which are not at right angles to one another[27]. Largest and most diverse group of sponges.
Glass Sponges (Hexactinillids)	Skeletons made of silica with 6 rays. Thought to be the most primitive of the sponges. Many of their tissues are syncetia (large cellular masses formed by many cells combining together in large multinucleated bodies). Lack of some cell types (such as archaeocytes). No specific motility cells. Able to move via electrical conductivity systems[28].

Sponges are extremely ancient life forms. The earliest fossils have been identified or occurring in the Ediacaran era[29]. They achieved fantastic diversity during the Cambrian and Cretaceous eras, of which the fossil record is only poorly reflective[30]. In many eras, sponges were in fact responsible for most of the reef building activity, despite that ecological niche being primarily filled by Cnidarians today. The first known sponge is called Paleophragmodictya, and is dated from 650-543 million years ago[31].

A number of biological forms appear to have originated in the sponges. The contractile ability of sponges appears to function systematically, allowing the sponge the ability to, in effect, cough and remove obstructions from its pores[32]. Sponges were also described by Dr. Helan Jaworski as presenting precursors to both lymphatic and glandular tissues[33]. Jaworski also likens the skeletal structures of sponges to our own bones. Other structures, like the flagella, are still clearly used in our reproductive tracts, such as in sperm and in the flagella lining the fallopian tubes.

Ecological Role of Sponges:

Sponges have played a major role in the ecology of earth since their evolution. In many epochs, such as the Cambrian and the Cretaceous, sponges made up a majority of the reefs on earth[34]. In the contemporary ecosystem, sponges still play an important role. One sponge can filter approximately 20 000 times its own volume of water in a single day, filtering up to 90% of the bacteria present[35]. Sponges thus play an important role in cleansing and purifying water, and ensuring the resulting health of their ecosystems. Sponges are often some of the first organisms to be adversely affected by water pollution, and now serve as important indicators of the ecological health of their ecosystems. The more primitive glass sponges are not as efficient at filtering bacteria, and likely have a great reliance on filtering out bits of organic matter[36].

Sponges also serve as hosts to many other organisms. A large variety of bacteria are hosted within the bodies of sponges in symbiotic, pathogenic or parasitic modes[37]. Sponges host photosynthetic organisms and may end up producing more oxygen and organic matter than the sponge itself consumes[38]. Some sponges in fact have a symbiotic relationship with photosynthetic organisms, such as cyanobacteria, and thus derive their nutrition from these symbionts[39]. Sponges also host a number of other animals, such as small crabs and crustaceans. One of our proved sponges (*Euplectella*) in fact is venerated in East Asian cultures for providing homes for pairs of mating shrimp.

Sponges also have a complex chemical life, and produce a wide variety of biomolecules that serve defensive purposes, though it remains uncertain whether the sponges themselves or commensal microbes synthesize these compounds[40].

Some sponges are also carnivorous! A few members of the *Clardorhizidae* (desmosponges) have small spicules that attach to crustaceans. These crustaceans are then soon engulfed by cells and externally digested in a process resembling immune system functioning in high animals[41].

Sponges are found in almost all shallow aquatic habitats, including both freshwater and marine habitats and in hostile biomes such as the Antarctic and the Arctic oceans. Figure 6 shows the relative concentrations of species throughout the globe[42]:

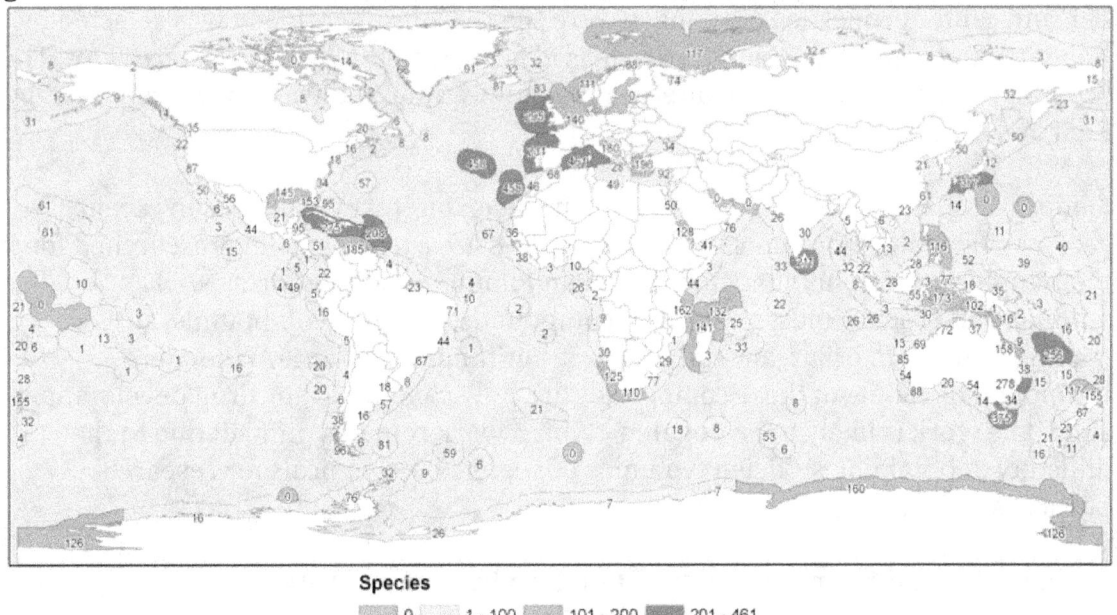

Figure 6: Concentration of Sponge Species Globally

Unique Characteristics of Sponges:

Sponges have a number of unique characteristics that interest us within homeopathy. In terms of their unique characteristics among animals we can examine:
1. Cells remain able to change forms and functions, in effect staying uncommitted to their roles. It is as if sponges are groups of single cells, rather than multicellular organisms.
2. Their lack of differentiation into separate tissues.
3. Their relative passivity compared to other animals.
4. Their ability to contract, which Jo Evans likens to coughing[43].
5. Sponges seem to form a basis for many of the structures of more complex animals, including glands, lungs, lymphatic tissue, reproductive systems and immune systems.
6. Their high concentrations of halogens, particularly iodine, further strengthen their connections to the glands, especially the thyroid.

History of Sponge Use in Homeopathy and Medicine:

Sponges have an extensive history of use within homeopathy, folk medicine and in contemporary allopathy.

Sponges have long been used in folk medicine for their effect on the glands. The sponge is traditionally toasted before use, and has been employed in disorders of hardening glands, bone issues and goiters[44]. It was also used to treat cancers[45].

The anti-goiter properties of sponges were the reasons the remedy was selected by Hahnemann for proving. Spongia toasta is described in the *Materia Medica Pura*. The proving was greatly expanded after Hahnemann's time, and was extensively used by Kent and others.

Modern allopathic medicine has also taken a keen interest in sponges in recent years. Many of the biochemicals within sponges are currently being researched for their applications in immunostimulation and immunomodulation as well as antimicrobial compounds, anticancer compounds, anti-HIV compounds, anti-malarials, anti-bacterials, anti-tuberculars, antifungals and anti-Alzheimers compounds [46]. Some of these compounds are now candidates for drug development, and some work is likely to be ongoing, though work remains difficult due to the difficulty of obtaining sufficient volumes of sponge biochemicals for research purposes[47].

Systematics and Characteristics of Porifera in Homeopathy:

Porifera are the earliest known animals, and fall within the first layer. This is confirmed by the available provings and cases. As layer one remedies, an emphasis on incarnational issues, existence or lack thereof and existential problems are to be expected within this family. Sponges also have large amounts of either calcium carbonate or silica in their skeletons, which will often lead to a Silica or Calcarea flavor to some of these remedies, but within an overall framework of layer one.

The sponges, based on the research and clinical cases thus far available, seem to have an emphasis on themes of <u>existing or not existing</u> depending on the outside circumstances, maintaining the <u>activities of daily life</u>, the <u>purpose and meaning of life</u> and <u>distortions of sensation</u>. In my work thus far I have found the sponges to have a <u>deep insecurity</u> about living and existing in the world. The nature of that insecurity, and the factors which particularly highlight it to the individual patient seem to be characteristic of the individual sponge remedy needed.

In terms of physical pictures, Sponges seem to share a common therapeutic affinity to the glandular system. Spongia has an incredible affinity to the pancreas, to such an extent that it has become one of the diagnostic remedies used by Dr. Jared Zeff, ND[48]. He administers it in a 200c potency for patients with suspected pancreatic issues, such as acute pancreatitis, noting an immediate decrease in pain within a few moments if the pancreas is in in a pathological condition. I have noticed a similar affinity to the thyroid gland in Badiaga, but have less clinical experience using this remedy diagnostically. This affinity requires more provings and clinical confirmations to be more firmly established, however.

As relatively few sponges have yet been proven, it is likely that our knowledge and understanding of this phylum will expand as more provings and clinical cases emerge. I do hope, in future editions and updates of this work, to have more sponge remedies to share with you.

As so few sponge remedies exist, it is difficult to differentiate the different evolutionary classes of sponges, as is possible in other animal phyla.

However, based on what we know about sponge evolution and morphology, some tentative themes may be proposed. I stress these themes remain <u>speculative</u> until a more extensive number of provings have been conducted. The themes are arranged from most primitive and basic to most well developed.

I will speculate also that the calcareous sponges and the desmonsponges may have more of a resemblance to the remedy Calacrea carbonicum, due to their calcium skeletons. I also will speculate that hexactinillid sponges will resemble silica in some respects, due to the silicate component of their skeletons.

Sponge Class	Possible themes
Hexactinillids (Glass Sponges)	The formation of syncetia (large multinucleated cells formed from the merging of smaller cells with single nuclei) suggests to me that the process of incarnation may be resisted in this group. They may wish to return and remerge with the divine. I suspect each individual remedy will have a particular issue that causes them to retreat back into the oneness out of which they emerged.
Desmosponges	The being is unsure of whether or not to commit to life. They perceive some of the necessary factors for being alive to be deficient, and thus avoid making a commitment and fully incarnating. Each individual remedy likely has a separate issue that stops them from fully committing to life.
Calcareous	The being feels ready to commit to life, but issues of inadequacy in some respect keep them from doing so. Each individual remedy likely has a separate inadequacy that prevents them from fully incarnating.

Below is a written text by Dr. Ghanshyam Kalathia on the Porifera from a sensation perspective. I would like to express my deep gratitude for this contribution from him, and I hope it serves to make the sponges more accessible to those homeopaths trained in a more sensation based approach.

The Sensation of Sponges- Dr. Ghanshyam Kalathia

No structure – formless
Sponges are multi-cellular (Metazoan) but very primitive animals. They are the first animals that evolved from unicellular organisms and who developed the capability to aggregate cells together and form a very basic structure where there are no connective materials between cell types. There are 11 types of cells that form the structure, but there is not any definite connection (tight bond) between different cell types. There are no tissues or organ systems, so only specific type of cells fulfill any specific function. There is no digestive system but digestion occurs; there is no connective tissue/skin but they have protection from outer world; there is no brain and nerve cells but locomotion occurs; there is no genital system but reproduction occurs. In short, sponges are metazoans but they work like protozoa (unicellular). The cell is everything because there is no system. The cells are there but there is no definite structure or form in them; there is no collective cooperation or coordination. Externally they have shape and some specific look, but systematically there is no structure or form. This aspect expresses in patients as having no form,

frameless, or having no structure. They are not able to explain what and why, there is feeling and sensation but they are not able to explain it. They are irresolute and unable to give out an adult description of themselves because they are very basic and simple. They are naïve. A sponge is someone who is unable to elaborate his experiences. There are frequent changes of topic within discussions, no surety etc.

- TALK, talking, talks; wandering– Spong
- AWKWARDNESS; work, about– Spong

No Borders – Uneven
In the sponges there is no symmetry or specific design in structure because they are just aggregations of cells, and all functions occur at the cellular level. There are not any connective channels between cells to carry food, body fluids, nutrients and other chemicals. Everything moves through the cell wall. So, this cell wall is the only important place where most of the vital functions are completed. Secondly, there is no skin or skin-related structure but some specialized cells called pinacocytes (plate-like cells) that give firm shape and structure to the whole sponge body. pinacocytes form a single-layered external skin over all other parts of the mesophyll (body fluid). Lack of a specialized border or sophisticated shape expresses in patients as feelings of having no border, no shape, mass-like, jelly-like or being uneven. The patient is extremely sensitive and even the slightest change in environment (house, family, support system, etc.) is enough to produce disaster. They take on others' emotions readily, as if everything is coming in and there is no border, (there is no defense). Because of this kind of extreme sensitivity the patient might be confused with a plant, such as a member of the Ranunculaceae. When we examine the case carefully, however, we can observe that case has multiple facets rather than an focus on one sensation and its exact and spontaneous opposite which would indicate a plant remedy.

Unstable – Changeable/Alternation
In the sponges there is not any structure to provide stability and firmness. Only pinacocytes give firm shape and structure to the whole sponge body. Pinacocytes are single-layered plate-like cells, so sponges have no multilayered skin as other evolved animals do. Pinacocytes are not tightly bonded with each other, so they can move and are able to make changes in their external appearance. In short, sponges have shape and structure but these are unstable and are changed at any time. sponges' instability and unsteadiness expresses in patients as changeability of symptoms and alteration of states. At one moment they are very happy and active, but the next moment they are sad, gloomy and lethargic or tired. Their talking style is also wandering from one topic to another; at one moment they are talking about a dream and at another moment they are talking about a skin complaint.

- ANGER; alternating with; cheerfulness– Spong
- ANXIETY; alternating with; cheerfulness/Alternating with; joy– Spong
- ANXIETY; periodical– Spong

- ANXIETY; sudden, paroxysmal; angina pectoris, in/ Heart complaints, in/ Throat– Spong
- CHEERFULNESS; alternating with; absence of mind/ Aversion to work/ Distraction/ Moroseness/ Quarrelsomeness/ Seriousness/ Vandalism– Spong
- IRRITABILITY; alternating with; cheerfulness– Spong
- MIRTH, hilarity, liveliness; alternating with; weeping/ Alternating with; irritability – Spong
- MOOD; changeable, variable– Spong
- SING, desires to; alternating with; distraction/ Alternating with; hatred of work– Spong
- VIVACIOUSNESS; alternating with; weeping, tearful mood– Spong
- WEEPING, tearful mood; alternating with; cheerfulness– Spong
- WEEPING, tearful mood; alternating with; irritability– Spong

Immature – Total Dependency
Sponges have a very basic structure, and their functional organs and systems (cells) are also very primitive. This aspect expresses in patients as immaturity. The person is very much child-like and underdeveloped. Their indecision, their changeability, their way of expression is enough for them to be judged as immature. They do not wish to make big changes and wish to hold onto that which is familiar. Because of their immaturity, they need others and so are very dependent. Their total dependency seeks the support from others who are stronger than them. Sometimes patients express this issue as incapability and there is no form and no structure, leading to the impression of a mineral remedy. In minerals this incapacity is a lack of something, while in sponges it is an incapacity for self-defense. The deeper experience is weakness, vulnerability, smallness, defenselessness, etc. Immaturity and dependency are more pronounced in non-evolved species like glass sponges, while evolved spaces like desmosponges, and calcareous and fresh water species are less immature and dependent.
- CONFIDENCE; want of self– Spong
- FOOLISH behavior; children, in– Spong
- JESTING; silly– Spong
- SUCCEEDS never– Spong
- TIMIDITY– Spong

Defenseless – Panicky/Fear of fear
We can easily understand that sponges are defenseless and without borders. They are not able to move and react like more sophisticated and developed animals. Their only defense is their spikes. The glass sponges have silica spikes, which are less

effective, and they are very much vulnerable. Demosponges have spongin and silicon dioxide spikes, which are a little more defensive than glass sponges. Calcareous sponges have tougher calcium carbonate spikes which offer the highest defense. Defenselessness is co-theme of 'without borders' and it expresses as extreme fear, anxiety, hypochondria and panic. The sponges main problem is that they are so irresolute that they hardly are able to discuss their fears or frights. The slightest change is too much for them, it creates turmoil and panic.

- ACTIVITY; palpitation, with– Badiaga
- ANTICIPATION; ailments from, agg– Spong
- ANXIETY; agg., ailments from– Spong
- ANXIETY; fainting, as from– Spong
- ANXIETY; fear, with– Spong
- ANXIETY; impatience, with– Spong
- ANXIETY; inconsolable– Spong
- ANXIETY; moaning, groaning, with– Spong
- ANXIETY; Pain, with; heart, in and about; perspiration, with/ Midnight; after; waking, on/ Bed; in; sit up, must/ Cramps, with/ Croup, in/ Driving him from place to place; dyspnea, in/ Heat; with; sudden/ Waking, on; dreams, from frightful– Spong
- ANXIETY; palpitation; with– Badiaga, Sponge
- ANXIETY; weariness of life, with– Spong
- DEATH; presentiment of – Spong
- DELUSIONS, imaginations; fire– Spong
- DREAMS; frightful, nightmare; midnight; after; four am– Badiaga
- DWELLS; events, on past disagreeable; frightful, mournful– Spong
- FEAR; death, of; suffocation, from/ dreams, of; terrible/ FEAR; suffocation, of; night– Spong
- FEAR; imaginary things– Badiaga
- FRIGHTENED easily; night/ Easily; perspiration, during/ Easily; waking; on– Spong
- INCONSOLABLE; heat, with, would rather die on the spot– Spong

Sensitivity
Sponges do not perceive the environment through senses and give appropriate reactions to it as plants do. Sponges are borderless, so they must be sensitive to perceive any threats. In sponges sensitivity is related to the need to perceive the environment for survival, and this additional sensitivity express in the following ways:

1. **Acute senses or active sixth sense:** Sponges do not have a nervous system at all, but they do have a very primary and non-developed method of perception for their surroundings. They sense environmental changes, they sense anything touching their external surface, and they sense the chemicals from seawater. Their main perception method is related to chemical signaling, similar to what occurs between different kinds of cells. They do not have a nervous system but they have neurotransmitters; they do not have a digestive system but they have enzymes; they do not have an endocrine system but they have hormones. So, their perception is different than other evolved animals. Even more, their entire survival is dependent on their sensual perceptions, so this issue expresses as higher sensitivity, exaggerated perceptions, the need to sense the surroundings, the need to sense the change inside, images of enlargement or unclear shapes, the need to sense the motion, sense the movement, sense the chemicals or smells, sense the clothes, awareness of body parts, awareness of internal organs, awareness of dimensions, and the sixth sense. It is not clairvoyance but almost a clairsentience. In short, we can presume that a sponge patient is much more sensual than others.
   - DELUSIONS, imaginations; body, body parts; enlarged– Badiaga
   - DELUSIONS, imaginations; enlarged; he is– Badiaga
   - DELUSIONS; clothing, clothes; uncomfortable/ Figures, sees/ Phantoms, sees; animated, lively / Hearing, of/ Motion; up and down, of– Spong
   - IRRITABILITY; noise, from– Badiaga
   - SENSES; acute– Badiaga
   - SENSITIVE, oversensitive; impressions, to all external/ Oversensitive; light, to/ Noise, to– Badiaga
   - SENSITIVE, oversensitive; impressions, to all external/ Sensual impressions, to– Spong
   - SENSUAL impressions, ailments from– Spong

2. **Touchy – Extremely sensitive/ Hyperesthesia:** Touch is very important for the sponges because they do not have any additional senses. They can perceive internal changes as well as outer changes on their cell walls. Sponge cells can perceive the chemicals, light, waves and vibrations and in this way they coordinate and make possible their routine work and defenses. This issue express as sensitivity to touch or hyperesthesia. This touchiness also expresses in patients as the feeling that everything overwhelming, or too much. They can't cope because they are vulnerable, weak, unprotected, small, and defenseless. They are so weak and defenseless that the challenge is the biggest turmoil for them. They are not able to carry on their daily routines. It

is as if they feel they are an alien or that they come from a different world. They feel they are totally unfit for this world.
- DELUSIONS, imaginations; touch, sensory– Spong
- SENSITIVE, oversensitive; touch, to– Spong

3. **Irritable, angry, rude, abrupt:** They are unfit and incapable, so the simplest thing is impossible for them and they feel as if anything is too much. Despite their incapacity they try to do things, but this creates turmoil in them. Their self-confidence is so low that the outcome is solely their reaction rather than the effect of their actions. They react with anger, aggression and irritability. Eventually these reactions express in an abrupt, abusive, capricious, defiant, contrary, rude, obstinate, quarrelsome attitude.
   - ABRUPT– Spong
   - ABUSIVE, insulting/ ANSWER, answering, answers; abruptly, shortly, curtly– Spong
   - ANGER, Ailment from agg– Badiaga
   - CAPRICIOUSNESS– Spong
   - CONTEMPTUOUS– Spong
   - CONTRADICT, disposition to– Spong
   - CONTRADICTION; intolerant of; children, in– Spong
   - CONTRARY– Spong
   - LOOKED at; cannot bear to be, agg– Spong
   - MALICIOUS, vindictive– Spong
   - MOROSE, sulky; talk, indisposed to, taciturn– Spong
   - OBSTINATE, headstrong; whooping cough, in– Spong
   - QUARRELSOMENESS, scolding; alternating with; gaiety and laughter– Spong
   - RUDENESS– Spong
   - WEEPING, tearful mood; anger, vexation; with– Spong

Reaction – No movement
They are not able to move or give active reactions as more evolved animals would, so they try to withdraw.
- BESIDE oneself, being; anxiety, after– Spong
- COMA vigil; sleeplessness, with– Spong
- CONCENTRATION; difficult; alternating with; gaiety and singing– Spong
- CONFUSION of mind; motion; agg/ Waking, on/ Walking; agg/ Intoxicated feeling– Spong
- DAY-DREAMING– Spong
- SENSES; dullness of, blunted; vertigo, in– Spong
- STUPEFACTION, as if intoxicated–Badiaga, Spong

Reaction – Contraction/Spasm vs. Expansion
Sponges are filter feeders and water flows from the inside out. For controlling water flow, they contract and expand their inlets and outlets. Sponges may also contract in order to reduce the area that is vulnerable to attack by predators, but as threats reduce they can resume their normal shape. So, contraction and expansion are basic survival mechanisms of sponges, and they express in the patient's language as contraction, retraction, becoming a ball, becoming smaller and then bigger again, tightening, etc. Superficially this can resemble the Cactaceae family but when we allow the case to go further and we examine all possible areas we can elicit many facets, which is common in the animal sensation pattern. In most cases contraction and expansion are present during the case-taking process.

Piercing, Shooting, pricking, throbbing sensation
For defense, sponges have spikes that can pierce if anything comes close to them. This issue expresses as piercing, shooting, darting, and pricking sensations in various places and parts of the body.

Marine theme
All marine animals share several common marine themes, and sponges are no exception:

1. **Marine theme – Hyperactive, cheerful, happy, joy, exhilaration, vivacious, excitable, impulsive, impatient**
    - ACTIVITY; hyperactive– Spong
    - ANGER; alternating with; cheerfulness/ alternating with; joy/– Spong
    - ANXIETY; impatience, with– Spong
    - CHEERFULNESS– Badiaga, Spong
    - CHEERFULNESS; alternating with; absence of mind/ Aversion to work/ Distraction/ Moroseness/ Quarrelsomeness/ Seriousness/ Vandalism– Spong
    - CHEERFULNESS; mischievous– Spong
    - CHEERFULNESS; prostration, before– Spong
    - CONCENTRATION; difficult; alternating with; gaiety and singing– Spong
    - DELIRIUM; cheerful, gay/ Fanciful/ Fantastic – Spong
    - DREAMS; carousing– Spong
    - EXCITEMENT, excitable; ailments from, agg– Badiaga, Spong
    - EXCITEMENT, excitable; morning– Spong
    - EXCITEMENT, excitable; nervous– Badiaga

- FANCIES; exaltation of; sleeplessness, with/ Sleep; falling asleep, on – Spong
- GREED, cupidity; eating, drinking, in– Spong
- HIGH-SPIRITED– Spong
- HURRY, haste; occupation, in; do several things at once, desires to– Spong
- HYSTERIA– Badiaga
- IDEAS; abundant; closing eyes– Spong
- IMPATIENCE– Badiaga, Spong
- JESTING; silly– Spong
- JOY; ailments from, agg.; excessive– Badiaga, Spong
- MEMORY; active– Badiaga
- MIRTH, hilarity, liveliness; alternating with; irritability– Spong
- MIRTH, hilarity, liveliness; alternating with; weeping– Spong
- MOOD; changeable, variable– Spong
- RESTLESSNESS, nervousness; children, in/ Anxious/ Drives one from place to place, must move/ Palpitation, during/ Respiratory complaints, with– Spong
- RESTLESSNESS, nervousness; night– Badiaga
- SING, desires to; alternating with; distraction/ Alternating with; hatred of work– Spong
- SING, desires to; perspiration, during/ Sadness; before/ Cheerful, joyously/ Involuntarily– Spong
- SUDDEN manifestations– Spong
- THOUGHTS; intrude and crowd around each other; closing eyes, a mass of subjects, some scientific, cross each other– Spong
- VIVACIOUSNESS; alternating with; weeping, tearful mood– Spong
- WEARY of life; inflammation of testes, in chronic– Spong
- WEEPING, tearful mood; alternating with; cheerfulness– Spong
- WEEPING, tearful mood; alternating with; vivacity– Spong
- WORK; mental; desire for– Badiaga

2. **Marine theme – Aversion to change, nostalgia of past**
   - SURPRISES agg– Badiaga

3. **Marine theme – Effect of Moon phases**
   - mind; MOON; full, during– Spong

Summary of the Sensation of Sponges:

According to Dr. Kalathia, the main sensations of Porifera can be summarized as:
- No Structure, Formless
- No Borders, Unevenness
- Instability, Change and Alternation
- Immaturity and Dependency
- Defenselessness, Panicky and Fearful
- Sensitivity
- Touchiness
- Primal sensation or Clairsentience
- Hyperesthesia
- Irritability, Rudeness and Abruptness

Their reactions to these basic sensations are:
- Lack of Motion
- Contraction and Spasm/Expansion
- Piercing, Shooting and Pricking

As well, common themes to all Marine Remedies are expressed within the Sponges.
- Exhilaration, hyperactivity, childlike joy
- Nostalgia
- A strong lunar effect

Again I thank Dr. Kalathia for this contribution, hoping more sensation-based practitioners find it useful.

Proven and Unproven Remedies within Porifera:

Unfortunately, very few sponges have yet been potentised, and fewer still have been proven. A summary is shown below. I would like to thank Jorg Wichmann for his animal systematics developed on his website, www.provings.info , which greatly simplified my research.

Sponge Class	Proven Remedies	Unproven Remedies
Glass(Hexactinellid) Sponges	Euplectella aspegillium (euple-a)[V]	None
Desmosponges	Spongia Toasta (spong), Badagia (bad)	Aplysina aerophoba (aply-a)[R], Chondosia reniformis (chond-r)[R], Tectitethya crypta (tect-c)[A]
Calcareous Sponges	None	Clathrina clathrus (clath-c)[R]
A- indicates the remedy is available from Ainsworth's R- indicates the remedy is available from Remedia V- indicates the remedy is available from Verfügbarkeit des Mittels bei Enzian Apotheke erfragen All other remedies are available at most pharmacies.		

Proving Suggestions:

Due to the extremely small numbers of sponges that have been homeopathically proven, any sponge provings would be very useful to the homeopathic world.

Due to the fact that two desmosponges have been proven, more extensive provings of both hexactinellid and calcareous sponges are strongly recommended. One remedy, Clathrina clathrus, has already been made into a remedy by Remedia, and so would be an excellent candidate for provings.

In terms of sponges that may be particularly fascinating to prove, I would like to suggest both the carnivorous sponges and photosynthetic sponges.

Carniverous sponges are desmosponges of the family cladorhizidae which passively capture small organisms and actively digest them extracellularly. I cannot even speculate on the homeopathic themes this sponge would have, but would be fascinated to see the results.

A second group of sponges that would be fascinating to prove are the photosynthetic sponges. This diverse group of sponges have evolved symbiotic relationships with photosynthetic algae or cyanobacteria, precursors of photosynthetic organelles of plants, bringing them very close to plants in their life modes[49]. This crossover, in the context of a layer one remedy family, would be fascinating to observe. These sponges are particularly abundant in Australia, so I do hope my Australian colleagues will take up a proving of these magnificent animals. One genus suggested would be the blue sponges, the genus *Collospongia*.

Lastly, while not strictly a sponge, another remedy that would be most illuminating is the choanoflagellates. This group of organisms represents the boundary between cells that live as individuals, and cells that begin to work together in organisms. They lie at a crucial point in evolutionary history, and thus they would have great significance within homeopathy. A proving of a pure culture of any choanoflagellate species would be most enlightening.

Materia Medica of Sponge Remedies[a]:

---

[a] For Ease of referencing:
  [M] refers to Murphy's Nature Materia Medica (2007). Accessed on Reference Works.
  [A] refers to Allen's Keynotes (1898). Accessed on Reference Works.

Badiaga (Bad, Spongilla lacustris)[50]:

Badiaga is a small freshwater desmosponge native to the northern hemisphere. Badiaga has been used extensively throughout homeopathy, but it's mental picture remains quite undeveloped. As such, I decided to undertake a trituration of this sponge to more thoroughly explore the mental image of this remedy. The mental image and key issues of this remedy derive mostly from my own trituration. My trituration notes are given upto C5 is given after this monograph.

Key Issues: The key issue is the <u>lack of the basic security needed to exist in the world</u>. They perceive the world as a dangerous and terrifying place and thus do not want to become fully incarnate within it. They will often either escape this danger by intellectualizing it, or through escaping to a dream realm of love and universal connection, the spirit realm where consciousness exists before its incarnation into the physical world.

Mind[M]: The mind picture of this remedy remains quite undeveloped, with only a few mind rubrics and no extensive mental/emotional picture yet developed, aside from my own trituration. The major symptoms identified in Murphy are a desire for mental work, and heart palpitations after pleasurable emotions. I identified a tendency to out of body escapism and intellectualizing away fears.

Main Physical Issues[M]: Muscular pain and soreness, worse motion and Worse motion of clothes[L]. Skin pain and soreness. Many issues with the lung, mucous flies out of the nostrils upon coughing. Corhyza. Pain in forehead that extends to eyeballs. Buboes and other syphilitic symptoms, such as neuralgias and bone pains. Swollen

Glands.

Clinical^M: Breast cancer, coryza, dandruff, rheumatism and muscular pain, syphilis, lungs and chronic coughs. Affinity to the thyroid gland. May be the Chronic of Opium.

Miasm: Syphilitic, Tuberculinic. Psoric

Color Preference: Black

Remedy Comparison: Aconite, Spongia, Euplectella, Syphilinum, Tuberculinum, Mercurius. Opium. Carcinosin.

Cases:

The following case was a long-term patient of mine. She was being treated for Euthyroid sick syndrome with iodine. Her daughter had recently been hit by a car (with no serious injury). The patient reported being severely affected by this. This feeling reminded her of some issues she experienced years earlier with her ex-husband during time in Guatemala together with their daughter, who was then quite young. A number of issues surrounding trauma and fear emerged in that period, of being unsafe, and a feeling of being unable to process or cope. Opium was prescribed, 200c one drop succussed daily. She reported some relief.

The patient later became ill with an acute gastrointestinal ailment. Carbo veg was prescribed for a bout of gassiness and feeling of bloating with some relief being reported from the prescription. At the following visit the patient reported that she felt her weight was out of control. She felt incredibly puffy and bloated. She felt spongey and almost blown up full of air. Previous issues of feeling unsafe with her ex-husband also re-emerged.

I prescribed a dry dose of Badiaga 10m.

One week later she experienced much more energy along with increased ability to do activities she was aspiring to do, but had previously lacked the energy for, such as exercising. She also reported a feeling of vertigo, unaffected by chiropractic treatment. She added that this feeling had also occurred during the period with her ex-husband in Guatemala. I changed the prescription to Badiaga 30c, 1 drop per day succussed. During subsequent follow-ups she noticed the feeling of vertigo having disappeared.

She was kept on Badiaga for two months, the potency being increased to 200c, and enjoyed a higher level of energy in respect to her earlier levels. Several months afterwards I changed her remedy to Carcinosin for suspected chronic viral infections with good success.

Trituration:

Badiaga: Why Do I Need to be Alive anyway?

This remedy was made from a mother tincture of Badiaga purchased from Helios. It was triturated by myself alone.

C1-3:

The first three levels of the trituration present a picture of a being beset by fear of existing. Existence seems uncertain and terrifying. There are no guarantees in life, and anything could potentially happen. On a conscious level the being copes with this fear by either rationalizing it away, or by escaping into an out-of-body dream state in which love is everywhere, and it has no separation between itself and that love. In both coping strategies, reality is so terrifying that the being cannot commit to it. In the words of the triturations itself "I am in life. I'm here in body. I just can't bring myself to commit to it due to fear and uncertainty"

C1:
Preformed September 15th 2014:
- I'm feeling a contraction of my aura. It's like my aura is shriveling up and drying out
    - It's as if my outer aura feels like a sponge. It is rough with a lot of holes in it
- My energy feels heavier and more childish
- I feel heaviness in my actions. Taking action seems to involve more effort than it normally does
- I feel slow and a bit stupid. I'm less reflective, less smart and much less suspicious. I am more accepting. I just take what comes to me, without a lot of processing
- I feel short of breath. I am a bit anxious because of losing my breath
- I feel a pressure on my chest, pushing down (i.e., a pressure on my anterior chest pushing dorsally)
    - This pressure is not hard or intense, but it is widespread
- I feel timid. I feel generally nervous, afraid and anxious
- I feel like there is a moisture in my lungs
    - The inside of my lungs feels damp
    - This dampness makes it harder to breathe

C2:
Preformed September 18th 2014:
- I feel really happy, but it is a false happiness

- o Deep down I am quite uncomfortable
- I feel as if I am wary or nervous about something, and the happiness is just a mask on the surface to conceal my deeper feelings
- Now I am afraid! I'm nervous. I'm not afraid of anything in particular, but I am just generally afraid
  - o Life in general seems quite scary
- My right knee hurts. I feel a soreness, stiffness, boring pain and heat [Note: this is a chronic symptom for me, which did not resolve greatly after this trituration]
- I'm really anxious. But nothing in particular is bothering me. It's a generalized anxiety
  - o I'm afraid
- I want to cover and hide and shrink away!
- I feel fear, and then a contraction inward. It is as if I just want to withdraw into myself and hide!
  - o I'm just so unsure and so insecure
  - o Everything is just so scary!
- I feel an odd feeling in my head
  - o I feel as if my brain is withdrawing from my meninges and stretching them inward
  - o The immediate inside of my skull feels as if it is being pulled in
    - It is like a pulsing pain in my head
  - o I also feel a kind of bilious feeling in my head as well
- My throat is a bit raw and sore
- I feel wary of my surroundings
- But alongside all of this fear there is a superficial happiness
  - o This happiness is almost like a mask, concealing how I really feel
- I look happy and pleasant, but deep down I am worried

C3:
Preformed September 19th 2014:
- I feel my neck and shoulder muscles tensing up
- I feel afraid again and again I have a superficial personality that is covering up this fear. Now (as opposed to C2) however this covering up is a bit more sophisticated
  - o I feel like I am intellectually justifying everything
  - o I'm using my intellect to reassure myself and to convince myself that I am not scared and that the world itself is not scary
- This fear is basic and primal. I am afraid of everything outside of myself
  - o It's as if I think the entire world is terrifying
  - o There is this basic existential insecurity. Life is scary!
    - And I can have two strategies
      - I can intellectually justify life to myself or I can wear a mask of happiness

- I'm afraid of everything. If something pleasing happens I'm less afraid (but not unafraid), but if something bad happens I am terrified!
    - Thinking of bad things makes me feel as if my insides are turning to ice!
- Existential insecurity: just existing is scary and dangerous!
    - Anything could happen!
- Anything could occur. You could get eaten! Anything could go wrong! There are no guarantees in life
    - How do you go about living when there is so much that could go wrong?
    - I feel so afraid that I am not really living life
    - <u>I am unable to commit to life</u>; it is so scary!
    - I look with fear on the whole enterprise of living
- I am having difficulty breathing through my nose; my nasal mucosa are swelling
- I keep getting these odd happy feelings. Mindless Happiness. I still feel the fear, but <u>it is like I am someplace else</u>. I'm outside my mind a bit. I'm experiencing a bit of dreamy escapism from the fear of life
    - It's as if I am in a very beautiful place. I see lots of colors, love is everywhere and I feel a lack of personal boundaries between me and this love I'm feeling. It's wonderful
        - I only feel this love in my head and upper body. My lower body still feels the fear
    - I am in life. I'm here in body. I just can't bring myself to commit to it due to fear and uncertainty.
- I feel dreamy again. It's so nice to go to that place
    - But I still feel that fear
        - It's like I am between the fear and the bliss of the beyond.
        - I'd just rather stay in the beyond. Why do I need to be alive anyway?

C4

The being now feels less of the fear that predominated in C1-3, along with an increased willingness to engage with life. With the fear gone, the being fully enters into their body and feels able to go about the world safely to accomplish something (they're not quite sure yet what that something is however). The being loves the world and what it encounters within it, in a simple and forthright way. No mention was made of the out-of-body dreamy state, and I do suspect that the being no longer feels the need to escape into that state any longer.

Preformed September 20th 2014:
- I feel a sense of peace and relaxation

- I feel inner relaxation. It is as if my insides no longer feel frozen (as they did in C3)
- Strange. Over the last day I have been attacked online a great deal, but it hasn't affected me as it often does
- I'm not as afraid. The fear has receded. Now I just feel an extraordinary exhaustion and fatigue in the absence of that fear
    - I feel sleepy, heavy and grounded as well
- I feel like I am here in my body for the first time. At first being here in a body was very exhausting, but now I feel very clear and levelheaded. I feel very lucid
- I feel as if I am no longer afraid of life!
    - I feel like my fear of what might happen and my fear of getting attacked kept me from fully committing to being alive
- Now I feel happy, lucid and ready to make my way in the world. I have no idea what I am going to do yet, but I am going to do something here!
- I feel happy with life. It definitely doesn't seem as scary as it once did
    - I can do it; I want to be here!
- Instead of fear I feel kindness toward the world. It's not a sophisticated feeling, but it is simple and humble and plain and beautiful. A simple feeling of loving being alive and love of the things I am encountering
    - I'm just experiencing life and the world. There is not a lot of complex thought about that process
- I feel simple, plain, unadorned. It is as if I am a very forthright peasant, or a healthy child
- I feel a flash of fear again which quickly recedes
- I feel a dull headache just behind my forehead
- I feel a bit of burning in the mid-back
    - Laterally from the spine to the curve of my ribs
    - This burning is on the surface of my skin

Dream: This dream occurred the night after the C4 trituration: I was homeless and sleeping outside comfortably, near a bus stop with my dogs. I was quite happy, chatting with everyone who walked by. I saw a big St. Bernard dog, and was under the impression that this dog was guarding me and watching over me, like some sort of spirit guide.

Normally the idea of being so insecure would have terrified me. But under the influence of Badagia C4, I could accept even homelessness, with the knowledge I would be happy and protected.

C5

This level of the trituration goes into the purposes a being has for coming into the physical body. A being has a set of goals they wish to accomplish. However, within the perspective of Badiaga C5, a great deal of obstacles could potentially prevent the being from reaching those goals. The being feels that it cannot trust the universe to offer what it needs to accomplish its goals. It feels torn between its desire to incarnate and reach those goals, and its fear of not accomplishing them. This leads to a state of being suspended between life and the pre life state described in C3, not fully being present in either state fully.

Preformed September 30th 2014:
- I feel a return of the fear from C3, but it is now more conscious, less primeval and less on the body level
    - It's almost like a mental wariness now.
- I'm less afraid as I am wary. I'm concerned about the dangers inherent to life. It's like there is a world of things out there that could go so wrong and interfere….. But interfere with what?
- Interfere with my life….. I feel like there are a lot of circumstances that can interfere with my life that lie outside of my control
- I feel like I have made a plan or a set of goals that I wish to accomplish over the course of my life
    - But I also feel like there is so much preventing me from accomplishing those goals. A million things lie outside of my control that could disrupt the entire process
- My life could be thrown off by any of an innumerable number of factors
    - How will I get my goals done? How can I trust in my life to do what I need it to?
    - I don't know. Right now I can't
- I cannot trust life. I can't trust existence to be what I expect it to be. I've got this basic existential insecurity about the properties of the universe itself. Will coming here (i.e. incarnating) work out? Will I do what I need to do here? What will happen if I don't?
- I don't know how to resolve this issue. And if I don't resolve it, I cannot move forward at all!
- I just shudder to think of everything that could go wrong
    - Why is life so cruel and so terrible? What is wrong with the universe that such terrible things happen and people get so off course?
- I'm scared to think of the reasons!
- I'm stuck between the fear of the issues that would make my life go wrong and the desire to incarnate and reach my goals. As a result, I stay between these two places, not really in either of them, unable to move forward. It's like I'm unable to decide whether to stay or go?
- This is the question. To be or not to be, to commit or not to commit, to love life or to fear it. Which way do I choose? Not making a choice is itself a choice, perhaps the worst one.
-

Trituration Level	Summary
C1-3	A basic existential insecurity. Life is scary and dangerous. The being copes with this either by intellectualizing or by escaping to an out-of-body state of love and universal connectedness. Physical life in a body is so scary that the being cannot commit to it!
C4	The fear of existing recedes and the being can fully commit to life. It fully enters its body and feel capable of going out into the world and make its way in the world.
C5	A being has certain purposes for incarnating. The being sees the Universe as a scary place which cannot be relied upon to provide that which is necessary to accomplish its goals. It becomes stuck between the desire to incarnate and its fear of doing so.

Euplectella aspergillum (euple-a)[51]:

Euplectella, known informally as Venus' Flower Basket, is a glass sponge native to deep ocean waters in East Asia. It is a well-known symbol of marital love in East Asian cultures because small shrimp live in opposite sex pairs within the sponge and raise families within it. The shrimp clean and provide metabolic wastes for the sponge, and the sponge attracts food for the shrimp, creating a true mutualism. This sponge was proven by a group of triturators in Berlin led by Heike Dahl, and published on their blog in German[52]. I have used Google Translate to translate the website and pdf of the proving into English, as no human English translation is yet available. The entirety of the summary section is presented below the main material medica. I have edited the wording of it so that it becomes more comprehensible in English, but cannot guarantee the complete accuracy of the Google translation, not being a German speaker myself. I eagerly await a better translation.

Key Issues: This glass sponge has the issue of <u>intellectual inadequacy</u>. The being feels <u>unable to cope with the mental aspects of life</u>, and thus it wishes to avoid incarnating completely and to remerge with the divine. The core problem is feeling completely <u>overwhelmed and wishing to leave life</u>.

Mind[53]:
Feeling of inadequacy
Feeling of excessive demand
Sense that I'm not doing it right, not perfect enough

Feel simple, low, stupid, stupid, inferior. I do not dare to say anything
Inferiority complexes
Fear of being laughed at because I do not do things so well (and also have the feeling that they could do it better)
I prefer to say nothing on the subject because of fear of feeling inadequate
Feeling of dullness, words do not come in the brain, words are omitted and not understood
Hectic, stressed
Stressed out: as if the schedule was full and there is no time to breathe
I feel like I am running out of time; everything goes too fast
Feeling one does not create what one will without a rushed connected feeling and breathlessness.
Time seems to fly by
Things are immediately forgotten (Similar feeling to Alzheimer's or dementia), it makes things easier-
I feel as if only mechanical work is worthwhile, and that I do not have to think
I have to act according to the instructions
I do not want to have to decide anything
"Always keep things beautifully simple, nothing complicated!" .... "Just ask!"
If you keep things simple, the stress is less
Complicated technical terms no longer make sense
Complicated sentences must be changed so that they are understood
Doing two things at once is not possible
Hunted feeling, everything seems to be messed up
Lost in thought, sitting there, stares to himself
Sluggish, tired brain
Feels naive in any case complex
Simple type of thinking
Desire to give the responsibility away to others
Image of a straight, simple tube
Dewdrop, drop shape

Main Physical Issues[54]:
Sensitivity generally increased. Sensitive to noise
Globus sensation in the throat
Hunted feeling settles on the breath, with tightness / heaviness in the chest
Lancinations
Sharp pain
Feeling of a dagger, knife in the body
Strong effect on the brain, memory loss
Blush, to have been caught not knowing something
Allergic symptoms (tearing eyes, nose tingle, scratchy throat)
Cold feelings
More feelings of weakness, overwhelmed
Stiff and uptight after getting up from the chair, better at incipient motion

Headaches during mental effort
Dullness during mental effort
Desire to be left alone

Clinical[b]: ADHD, autism, developmental disabilities, allergies, lung and chest issues, alzheimers, dementia.

Miasm: ?

Color Preference: ?

Remedy Comparison: Baryta carb, Lepisma saccharina, Alumina, Graphites, Physalia phyalis. Crustacean remedies. Other sponges. Other layer one remedies.

---

[b] All suggested clinical uses are speculative in Euplectella, based on the Proving.

Spongia toasta (Spong, Euspongia officinalis)[55]:

Spongia toasta was one of the original remedies proved by Hahnemann and given in his Materia Medica Pura. The remedy came to his attention due to its use as a folk medicine for the treatment of goiter[56]. This remedy has been extensively used in acute and chronic disease throughout the history of homeopathy. Recently its mental picture has been greatly expanded through the work of several key authors, such as Irene Shlingensiepen-Brysch[57] and Liz Lalor[58].

Key Issues: The Key issue with Spongia is the <u>need for a secure social environment in which they can incarnate</u>. Spongia feels as if they <u>do not exist, or are not incarnate without their social interactions</u>. However, they are not sure if they can count on their social contacts to consistently be there to support them. This creates a hesitancy to fully incarnate and commit to life. This can also lead to acceptance of a great deal of abuse, in order to maintain their social network.

Mind[M]: Anxiety and difficulty breathing, fear of suffocation, fear of the future. Liz Lalor emphasizes the social difficulties Spongia has, as well as difficulties with trauma, fear of exposure of their inner weakness, and their ability to see things as if through the eyes of others[59].

Main Physical Issues[M]: Heart valve issues leading to heart hypertrophy. Dry cough of chronic heart disease. Glandular issues, especially thyroid. Dry mucous membranes. Coughing issues, worse with dry wind. Acute infections, such as diphtheria, whooping cough and croup, as well as other more chronic coughs. Characteristic cough sounding like the sawing of a pine board, worse at midnight. Feels a plug in larynx. Anxiety attacks with heat.

Clinical[M]: Coughs, diptheria, whooping cough, croup, heart issues, heart hypertrophy, thyroid disorders, goiter, orchitis, asthma, exhaustion, bullying, PTSD[60]. Suggested by Jared Zeff to have an affinity to the Pancreas.

Miasm: Tuberculinic[A], Psoric

Color Preference: 15-16E

Remedy Comparisson: Tuberculinum, Aconite, Saccharum offinarum. Other sponges.

Cases:

Patient presents April 4th 2014 with a desire to get her homeopathic case taken.

Her major issue is not knowing her own personality. She lives with others, but when she is alone, she feels nothing and does nothing. When asked to elaborate she states that without the company of others she feels like there is nothing happening; no thought, no feeling, nothing at all. She often will spend entire weekends in bed. She feels removed from what is occurring, and doesn't know what is happening around her.

When she is with others, she feels a distinct change. She feels happy and very grateful for her life. She has the distinct sensation that she has no right to feel anything besides good feelings, emotions and gratitude.

When prompted to describe her feeling of gratitude, she describes a feeling of fullness or bigness, feeling so full that she feels like things are spilling out of her, like sunlight shooting out of her chest region. She feels fuller, taller, and says she feels able to glide or float through the world. She describes herself as able to enjoy the simple things in life, and to take pleasure in her day-to-day activities.

She then gets into a distinct source sensation, describing a sort of structure around her. It is a spherical structure with a green tinted edge. It is porous, with a number of small dimples on the outside that do not penetrate to the middle, where she perceives herself. It does not move.

In her case, we see a basic state of flatness and nonactivity, characteristic of the sponges, combined with a sense of fullness associated with social activity, and taking pleasure in the simple acts of daily living. In my interpretation she was coming out of a state of nondifferentiation and into life, taking pleasure in social interactions and activities of daily living. She however was not yet able to sustain this outside of social interaction, hence the inability to sustain this feeling while by herself.

The remedy prescribed was Spongia 1M I drop per day, succussed with each dose.

We followed up on May 2, 2014. The patient reported feeling very introspective. She realized that she didn't know how to trust others, and that she did not consciously know how to be herself. She reported being very emotional during her menses, and having long discussions with her partner about her lack of trust.

Trust to her was a surrender to someone, being able to fully count on them being there.

The patient also was much more assertive with others who were disrespectful towards her. She was able to voice her concerns, rather than simply internalizing them as she had done previously.

When taken into her source again, she reported that the borders of the spongelike object were more blurry than before, and its interaction with the outer world had grown misty. The patient also reported feeling a lot better in general for the first week of taking the remedy, but this had stopped soon afterwards.

The emergence of her trust and identity issues into the conscious mind was a very good sign that she was exteriorizing her psychic conflicts (unconscious to conscious usually is exteriorization in my experience). The assertiveness and less defined source were also signs that we were close. The presence of the same issues, and the lack of a feeling of resolution of the state, led me to suspect that we needed a higher potency.

I prescribed Pascoefem™, 10 drops per day to deal with the PMS, and spongia 10m, 1 drop per day, sucussed each time.

At her next follow-up the patient reported feeling better still. She reported a tonsil infection in the last month, identical to those she had at 12 years old. It last 2-3 days, then resolved spontaneously. Her relationship with her boyfriend had improved, and she was now excited about planning her future together with him.

Also, for the first time, the patient reported a history of binge eating. She would periodically eat a large amount of junk food and feel terrible afterwards. She had engaged with a counselor to help her address these issues. Overall she felt wonderful. Her biggest problem was containing her joy at being alive. She did report some residual feelings of her state.

I was very pleased with the progress of her social relationships, that she both felt comfortable in herself enough to tell me about her eating disorder, and that she was healthy enough to begin to address it without external prompts. I had her shift her remedy upward to 50M, one succussed drop taken daily.

This patient later received Black Hole, Oak, Rainbow and is currently on Deer tick.

-Dr. Paul Theriault, BSc, ND

Further cases:

Two further cases of sponge are available in other publications. A beautiful case of spongia using the source method, and almost exactly paralleling the language used in Case 1, is available in Irene Schlingensiepen-Brysch's work "The Source in Homeopathy"[61]. This beautiful case is highly recommended. Irene's case, followed over several years, emphasizes the patient's need for structure, for another person to structure her life around, and the resulting grief she feels when this person becomes unavailable.

A second excellent case was published by Liz Lalor in Links[62]. In this case, a woman presents with issues of not being synchronized with the social behaviors of others, and finding others' social perceptions infiltrating her own psyche.

Summary of Key Issues of Sponges:

Remedy	Main Issue within the Table
Badiaga	I cannot incarnate because the world is a terrible place, full of danger and insecurity. I lack the security I need to incarnate.
Euplectella aspergilum	I do not wish to incarnate because I cannot deal with the mental effort required in life.
Spongia Toasta	I lack the social support I need to exist.

Cnidarians:

Cnidarians are a phyla of animals defined by the presence of cnidocytes, or stinging cells, by radial or circular symmetry and by the presence of planula larvae at some point in the life cycle[63]. Cnidarians are a diverse phylum characterized by wide variation in form, but all of which share a basic body plan. They include such creatures as jellyfish, anemones, corals and sea pens. Cnidarians are exclusively aquatic, with some members able to live in fresh water.

Like the sponges and the ctenophores, the Cnidarians have a basic body plan consisting of a layer of internal cells (gastrodermis) and one of external cells (the epidermis) united by a gelatinous layer (the mesoglea) between them. Cells possess linkages to one another in the form of basement membranes and are fully committed to their specialized roles, unlike those in porifera[64]. Specialized cells known as cnidocytes are specialized cells, unique to this phyla, which shoot out small harpoon-like barbs that may be specialized for gripping prey, delivering toxins, or wrapping around small objects[65].

Cnidarians possess both nervous and digestive systems. The digestive system uses one orifice for both intake of food and expulsion of waste. The nervous systems lack centralization, but have widely diffuse and responsive neural nets that perceive and respond to stimuli in a decentralized way[66] Some cnidarians, such as the cubozoans, have eyes that are extremely well developed[67]. Cnidarians also possess true muscles, coordinated by their nervous systems, allowing for whole organism movement[68]. Most cnidarians possess structures called tentacles that serve as structures for the cnidarian to manipulate its environment. Cnidarians lack circulatory systems, allowing nutrients and gases to diffuse.

Cnidarians possess two basic body plans. The first (and arguably oldest) form is that of the polyp. The polyp possesses a base (the aboral end) which attaches to a solid service. The oral end with tentacles and the mouth/anus opens to the water, from which the cnidarian derives food. All cnidarians possess polyp forms, suggesting that this form is the most ancient of the two. The second form (which is not universal within the Cnidarians) is the medusa. The medusa form is free floating in the water, and uses muscular contraction to actively move itself. Most of the scyphozoa, cubozoa and hydrozoa (free living jellyfish) have medusa stages, while the anthozoa (corals and anemones) do not.

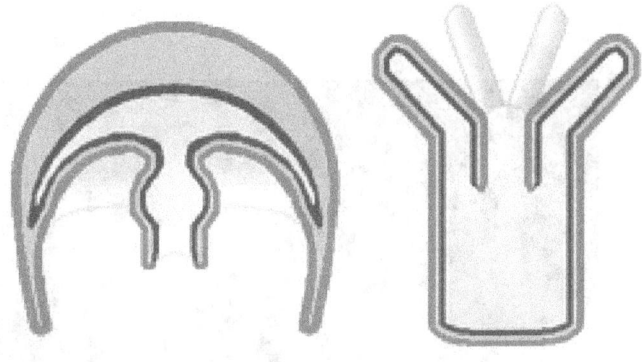

Medusa body plan is on the left, polyp on the right.
- Mesoglea
- Epidermis
- Gastroderm
- Digestive cavity

Figure 3: Cnidarian Body Plans[69]

Cnidarians can reproduce both sexually and asexually. All species can reproduce by budding, and all are capable of sexual reproduction. Individuals that bud will emerge from their parent organism at the stage of development that their parent was in at the time of budding (medusa for instance), while all sexual reproduction creates polyp forms. These polyps develop into medusa, as shown in the diagram below[70]:

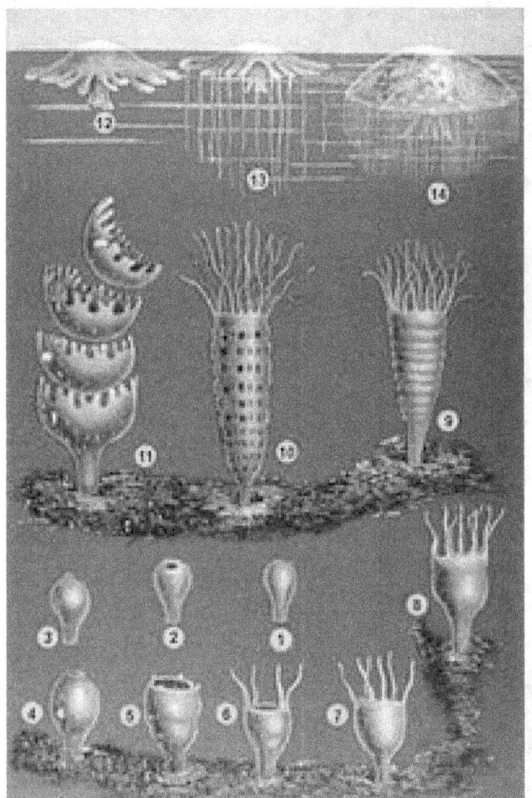

Cnidarians have a variety of ways of feeding. Many species are carnivorous. They use their cnidocytes to sting prey, and their tentacles to move it towards their mouth for ingestion. Some species, such as many corals, are commensal with algae. Algae living within the tentacles of the individual coral polyps produce nutrients and sugars for the coral and also assist in the fixation of calcium carbonate for the coral skeletons[71]. Some coral species are in fact completely dependent upon algae for their nutrition.

Cnidarians have several different classes, each of which has its own unique characteristics, and variations on the basic body plan. They are called the Anthozoa, the Hydrozoa, Scyphozoa, Cubozoa and the recently added Staurozoa.

Figure 4: Medusozoan development

The Anthozoa are composed of the corals and anenomes. They are defined by the absence of a medusa stage, circular mitochondrial DNA (which is shared with more advanced animals) and several other physiological features[72]. There are about 6000 species, showing a great deal of diversity, and a considerable fossil record. Anthozoa show both colonial and solitary forms, and they can generate external frameworks out of calcium carbonate[73]. This is of incredible ecological importance, with corals forming the basis of coral reefs, some of the most ecologically diverse biomes on the planet. [74]

Figure 5; Orange Sea Pen and Anemones[12]

Figure 6: *Aequorea victoria*, a Hydrozoan[14]

The Hydrozoa are a subphyla with a generally greater medusa representation, and several other physiological features held in common[75]. This phylum is remarkable for including colonial organisms which include both medusa and polyp forms, such as in *Physalia physalis*. It also includes a number of organisms which, while resembling corals and anenomes, are not members of the anthozoa[76].

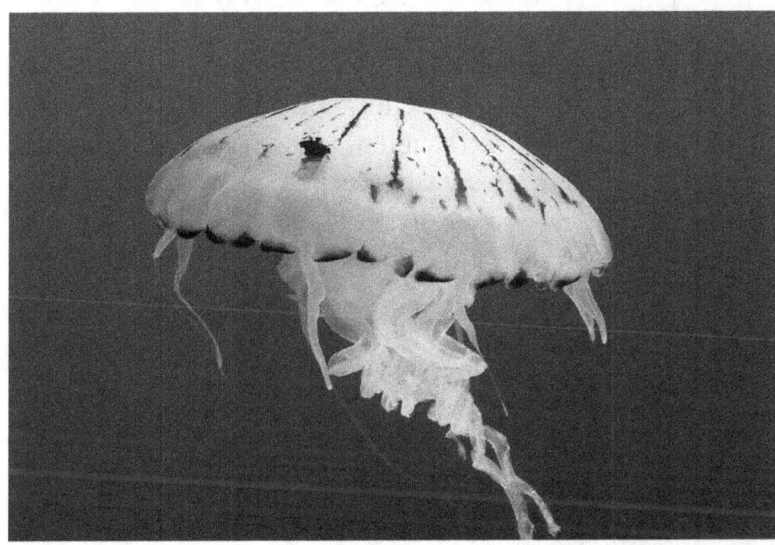

Figure 7: *Chrysaora colorata*, a Scyphozoan[17]

The Scyphozoa, or true jellyfish are defined by engaging in a process of specialized asexual reproduction[77]. They are active swimmers, moving water by contracting their bodies. They possess relatively well developed digestive systems, and while neurologically simple, they can display complex behaviors such as daily migrations[78], [79].

The cubozoa are defined by their cubelike bodies, and their relatively complex eyes[80]. They bear tentacles in groups of four, and they tend to contract their bells differently than the Scyphozoa, leading to faster propulsion[81]. This is the smallest group of the Cnidaria, but its members tend to be very venomous, with several human deaths occurring annually[82].

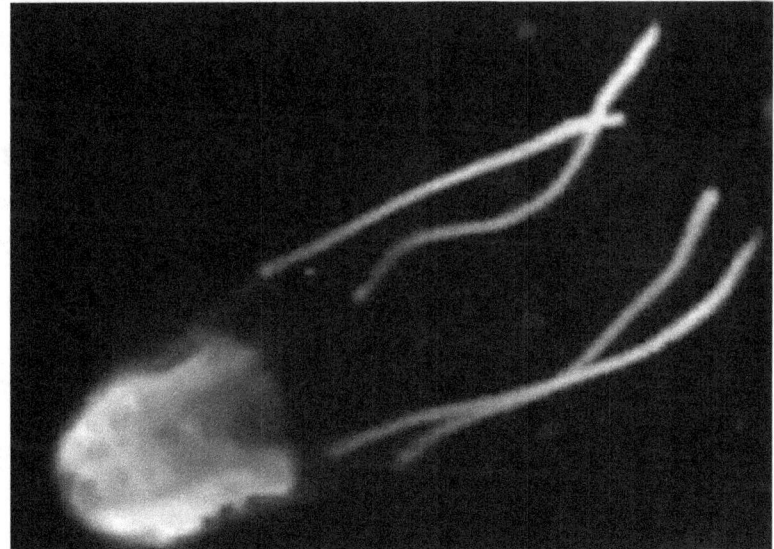

Figure 8: *Curukia barnesi*, a Cubozoan[20]

The final, and most recently added class is the Staurozoa. They do not alternate between medusa and polyps, but live their lives as attached medusa. They live attached to a surface, and use their tenacles to capture prey. An example is shown left, of Haliclystus stejnegeri[83].

Figure 9: *Haliclystus stejnegeri*

The classes are summarized in the table below.

Class	Description
Anthozoa	Lack medusa stage. Likely most closely related to higher animals due to the circular DNA in the mitochondria (other cnidarians have linear mitochondrial DNA[84]). Includes anemones, coral and sea pens. Usually non-mobile, or very slow moving. Can be both solitary and colonial.
Hydrozoa	Medusa in adult form, but having a greater representation of the polyp stage than other subphyla. Includes colonial forms with both medusa and polyp forms of the same species coexisting cooperatively, and single solitary forms.
Scyphozoa	Known as the true jellyfish. Active swimmers able to move via contraction of their cup shaped bodies. Complex internal digestive structures. Includes the largest jellyfish (2m across). Contains some light sensitive cells. Found in all oceans.[85]
Cubozoa	Primarily medusal. Cup of medusa stage is cube shaped (hence the name). Very poisonous. Restricted to tropical and subtropical oceans[86]
Staurozoa	Recently recognized as a separate subphyla. These organisms resemble anthozoa. Live as medusa attached to a surface. Live in colder waters[87]. No homeopathic representation.

One alternate classification divides the cnidarians between the anthozoa and the Medusuzoa (jellyfish). Due to the lack of multiple provings of medusozoa, this division will be adopted within the table of animals, until further information becomes available. The division of the Cnidarians into 5 classes, rather than two, however is likely to be more biologically accurate.

Ecology of Cnidarians

Cnidarians play immense ecological roles. Corals play an incredibly important role in oceanic ecosystems. This is especially true in many tropical regions with otherwise low levels of phytoplankton. The plankton symbiosis within corals forms the basis for the incredibly productive ecosystems of coral reefs, and provides the basis for many other creatures such as fish, arthropods and many sea mammals. Reefs are among the largest and most productive ecosystems in the world. The famous great barrier reef stretches for over 900 kilometers. Corals are preyed on by the crown of thorns starfish, butterfly fish and parrot fish.

Other cnidarians play important roles as predators and prey. Jellyfish form a major component of the diet of many creatures, such as sea turtles and sea slugs.

In light of the increasing concentrations of carbon dioxide, it has been discovered that many corals are less able to fix calcium carbonate to form their skeletons. The long term implications of this process are not yet known, but it is speculated that it may slow down reef formation, upsetting the sometimes delicate balance between reef formation and destruction.

The recent trend towards more acidic oceans and overfishing has also increased the population of jellyfish as well, again with unknown ecological consequences.

Evolution of Cnidarians

Cnidarians are an ancient evolutionary lineage; originating in the Ediacaran era when coral fossils are first identified[88]. Due to the lack of fossils of medusa, their fossil record is considerably more sparse, leading to considerable ambiguity in the understanding of their evolutionary development. The relationship of the pure polyp cnidarians (such as corals) and the medusal cnidarians still is difficult to ascertain.

Genomic evidence has however clearly pointed to the calcareous sponges as being more closely related to cnidarians, rather than to other sponges[89]. It can be hypothesized that the calcareous skeletons of the calcareous sponges evolved into the stony bodies of corals. Another hypothesis is that calcareous sponges evolved into an anemone like solitary creature, from which then evolved the colonial structures and skeletons of most corals.

The exact relationship between cnidarians and the bilaterally symmetrical animals (bilatera- essentially most other animal phyla) remains unclear. It is likely that anthozoa, with their circular mitochondria genome (shared with bilatera) are the most closely related to bilatera.

Unique Characteristics of Cnidarians:

1. Cnidarian cells are **fully committed to their roles**, losing the totipotency (ability for fully differentiated cells to change roles) that sponge cells have.
2. The first differentiation of cells into **separate tissues.**
3. The ability to actively **move the whole organism quickly**.
4. The development of the **peripheral nervous system, without a central nervous system.**
5. Many **early sensory structures**.
6. Soft **unprotected bodies**, which can easily be destroyed by predators or an unfavorable environment.
7. The presences of stinging cells, called cnidocytes

Human Uses of Cnidarians in Food and Medicine:

Humans have made extensive use of cnidarians over the ages. Jellyfish have been consumed in East Asian cultures for centuries, and a significant commercial fisheries exists to this day in both Asia and the United States[90]. The most commercially important species is *Stomolophus meleagris which has been suggested to have aniinflammatory effects when consumed*[91].

Coral have also been used extensively throughout the ages, both medicinally and decoratively. They were used in the middle ages to prevent epilepsy, hemorrhages, gonorrhea, leucorrhea, looseness of the teeth, kidney and urinary bladder afflictions, depression and issues of the liver and spleen[92]. Coral have also been used extensively in crystal healing and in Tibetan, Chinese and Ayurvedic medicine[93].

Bleached coral has also been used within allopathic medicine as a substrate for bone grafts, due to its unique ability to be replaced by human bone regrowth over time[94]. Various species have also been used as sources of antiviral, antifungal, antibacterial and anti-inflammatory substances[95].

Cnidarians have been proven very early on in homeopathy. Both Corallium rubrum and Medusa, or Aurelia aurita, were proven in the 19th century and are included within Allen's Encyclopedia[96]. Though neither have become polycrests, both have been widely used. Other cnidarians were proven in the 1990s and later.

Systematics and Characteristics of Cnidarians in Homeopathy[cc]:

Cnidarians are some of the earliest creatures, but they show the fundamentals of many more advanced structures of other animals. Based on the existing provings and cases, we can determine that the Cnidaria fall within layer two. As such, issues of boundaries, separateness, and leaving the divine for the individual tend to predominate.

Cnidarians appear to <u>not have yet built the boundary</u> separating the being from the outside world. As such, cnidarians have extreme <u>exposure to their outside environment</u>, they <u>cannot effectively separate themselves from their outer circumstances and outer environment.</u> The cnidarian problem has to do with the effect of this exposure on themselves. As layer two remedies, their selves are as yet incompletely developed. This incomplete personal development means that they are incapable of maintaining their sense of selves in the face of external stimuli.

Their massive exposure to their environment has the effect of <u>destroying the integrity of their selves.</u> Their <u>sense of self literally dissolves upon too intense exposure to the outside world</u>, leaving them extraordinarily dependent upon their environment.

Each cnidarian has a particular factor within it's environment that especially affects them. This factor will disrupt their sense of self, and is the key to finding the exact remedy the patient needs.

The two divisions of the cnidarians have different responses to this set of conditions. The <u>anthozoa experience this overload and withdraw from it</u>. In the corals, this is modified slightly into withdrawal into a small sheltered environment, not unlike the hollow spaces in which corals live. The main issue in the anthozoa is typically related to being unable to withdraw from these stimuli, causing the disruption of self. The <u>medusozoa are unable to withdraw and so need to exert an effort to cope with the external world</u>[97]. If the external world is beyond what the individual can cope with, then their self is disrupted.

Cnidarian Class	Summary
Anthozoa	Exposure to the external world destroys the sense of self. The being withdraws from the external world to cope.
Medusoza	Exposure to the external world destroys the sense of self. The being needs to expend an effort to cope with the external world.

[cc] I would like to thank Jo Evans and Dinesh Chauhan for their published works on the Cnidarians. Both contributed immensely towards my understanding of this phyla

Below is a summary of the themes of the Cnidarians written by Dr Ghanshyam Kalathia, edited for clairity. I am most grateful for his more sensation based approach and for his contributions.

Sensations of Cnidarians- Dr. Ghanshyam Kalathia

Little Structure – Jelly-like [Have Form]
Cnidarians are more advanced, developed and evolved than sponges because their cells are not just aggregating, but instead they form simple kinds of tissue. Here the structure is beginning to take form. There is structure but it is very simple and it has an undefined form. This expresses in patients as a feeling of little form, jelly-like, balloon like, umbrella like, vacuum, empty, tube-like. During a person's narration we easily presume that person knows who he is and what he wishes to talk about, but the speech of jellyfish patients has hesitation, as if their speech is jiggling. Their speech is trembling and they have little confidence in their own narration. They repeat some words several times, but with several mistakes. They are able to speak rather than become speechless as sponge cases do. Their talk and working style are still not precise and they make a lot of mistakes.
- ANXIETY; speech, with difficult– Medus
- TALK, talking, talks; foolish– Chir-f
- MISTAKES, making; talking, in; words, wrong– Stoi-k

These are the animals that have for first time succeeded in taking form, so this form is very important for them. Cnidarian patients are very aware and conscious about their appearance. They look a thousand times in the mirror and if there is any scar or pimple on the face they will get scared and behave as if they were diagnosed with cancer of the skin! They are also conscious about what they gain/achieve, because they have created something from nothing (nothing is associated with the Porifera). Whatever they achieve is very important for them and they are very much excited about it. They show off their achievements to everybody and are very possessive of them. They do not wish to lose it. So, in this way, they look childish.
Expression words: *Little form, Jelly-like, Balloon like, Umbrella like, Vacuum, Empty, Tube like, Cylinder, Jiggling, Jet propulsion, Hesitating speech, Trembling Speech, Repeat the word several times, Not so precise, Conscious about appearance, Conscious about your own body, Recognize difference between your own form and others*

Thin Borders – Anything Can Enter Inside of You
In the cnidarian body there are two cell layers; an outer epidermis or ectoderm, and an inner gastrodermis or endoderm. These two layers are not complex structures, just simple layers of single cells. They have a border but it is newly formed and very thin. It is still not enough to protect them and not defensive enough to be tough. The patient is very sensitive and gets affected by the smallest thing. Cnidarian patients are so sensitive that they get scared from even thinking of touch (pain, suffering, insult, vexation etc.) that they can react suddenly, impulsively and exaggeratedly. (cnidarians are more sensitive and reactive than porifera)

Expression words: *Thin border, Thin skinned, Very sensitive, Not tough enough, Permeability, Weak, Fragile, Can burst open with smallest touch, Bubble like, Smallest touch is unbearable, Reactive, Impulsive, Sensitive to pain, Fear of suffering, Fear of injury, Get affected by insult and vexation*

Unstable – Jiggling Like Jelly

Cnidarians are able to take a form so they have a little stability, but it is not long lasting stability. It is temporary stability, and they have extreme fear of losing it. With form they have awareness of their own being, but they are still very weak, vulnerable and fragile. Their stability is not enough so they jiggle when they encounter the minutest difficulties. To deal with this, they will often seek support. Their instability also expresses in the form of awkwardness, confusion and their immaturity in behaving and reacting.

- AWKWARDNESS; knocking against things/ Accident prone – Chir-f, Stoi-k
- CONFUSION of mind; identity, as to his; sexual– Medus
- DELUSIONS, imaginations; woman, he is a– Medus
- MISTAKES, making; localities, in/ Perception, of/ Writing, in; words – Stoi-k
- TIME; loss of conception of– Stoi-k

Because of their weak structure and instability, they feel they are jiggling and need support. Their development is up to the level of a small child or toddler, and they need support as a small child does. They need support to be able to stand on their own feet and to learn how to keep balance. They need support to be able to talk, walk and run. They need support to feed. In sponges the support is there and is enough for them while here they need support to be in an active role. They are needy, demanding, complaining and wanting lots of attention from their support system. They are very possessive of their support system. So, in short, without support they are completely disabled, unbalanced and vulnerable. They need support every moment in their lives.

- CONFIDENCE; want of self; failure, he is a/ Inadequate, feels/ Support, desires – Chir-f, Stoi-k
- CONFUSION of mind; concentrate the mind, on attempting to/ Daily affairs, about– Chir-f
- DELUSIONS, imaginations; disabled, being/ Protection, defense, has no – Stoi-k
- MOOD; changeable, variable– Chir-f
- SENSES; confused– Chir-f

Expression words: *Awkward, Making mistakes, Instable, Unbalanced, Not able to keep balance, Just start to learn balancing, Tottering, Faltering, Stumbling, Staggering, Wobbling, Shaking, Need support as active role, Needy, Demanding, Complaining, Want lot of attention, Possessive*

Panicky/ Frightful but Defensive
The little structure and stability that cnidarians have is not enough for them. They get scared and panicked if they are on their own. They do not accept instability as passively as the porifera do. They try to react after being frightened, they get nervous or excited. Their entire system gets involved in reactions so they can display a panicky state with a lot of anxiety, excitement and restlessness.
- ANXIETY; sleep; before/ Tranquility, with/ Stomach, in/ Causeless– Chir-f
- ANXIETY; speech, with difficult– Medus
- DREAMS; anxious/ Frightful, nightmare – Cor-r
- EXCITEMENT, excitable; ailments from, agg– Cor-r
- EXCITEMENT, excitable; amel.; mental and emotional amelioration– Cor-r
- IMPATIENCE– Stoi-k
- RESTLESSNESS, nervousness; busy– Chir-f
- SUDDEN manifestations– Chir-f

Expression words: *Scared easily, Frighten easily, Anxious nervously, Anxiety with excitement, Fear with restlessness, Impatience, Need immediate attention and assurance for safety*

Extreme Sensitivity
Cnidarians' activities are coordinated by a decentralized nerve net and simple sensory receptors. They are the first animals who evolved with a very simple nervous system, and they are the first animals able to take stimuli and react appropriately. Being able to react is their unique defense mechanism. Every individual medicine has their unique reacting pattern.
- NOISE; agg– Chir-f, (STARTING, startled; noise, from) – Stoi-k
- SENSITIVE, oversensitive; impressions, to all external/ Sensual impressions, to– Chir-f
- SENSUAL impressions, ailments from– Chir-f
- SENSES; acute; body, of; pleasant– Stoi-k
- SENSITIVE, oversensitive; impressions, to all external/ Sensual impressions, to – Stoi-k

Cnidarians have a well defined sensitivity rather than the undefined sensitivity of Porifera. They have nerves and so are extremely sensitive to pain, suffering and injury. At a mental level they are sensitive to insult and vexation. Porifera are not able to feel beyond touch, so their focus of sensitivity is touch and mentally they can display quite a bit of touchiness.
- DELUSIONS, imaginations; injury; injured, of being– Stoi-k
- DELUSIONS, imaginations; insulted, he or she is– Stoi-k
- DREAMS; accidents, of; crash of airplane– Stoi-k

- DREAMS; mortification, humiliation– Stoi-k
- FEAR; pain, of/ Suffering, of / PAIN; agg– Cor-r
- IRRITABILITY; Pain, during– Chir-f
- LAMENTING, bemoaning, wailing; pain, from– Cor-r
- QUARRELSOMENESS, scolding; pains; during– Cor-r
- SHRIEKING, screaming, shouting; pain, during– Chir-f

(1) Acute senses

Cnidarians are the first animals that are able to sense their surroundings in reality. It is as if someone who has been blind is now suddenly able to see. When, for the first time he is able to see, what is his feeling and reaction? This is exactly the situation with cnidarians. They are very excited and so try to touch everything, sense everything, experience everything as if they wish to perceive the entire world. During consultations patients describe everything with very minutest detail, as if they are sensing thier surroundings at present moment. For example they say that everything is so beautiful, so pleasant and so colorful. This is the reason they have artistic aptitude, they love everything and they wish to explore everything that is new to them. They like every pleasurable sound. They like singing, music and dance. They like all physical rhythmic activity. Their artistic aptitude is due to sensuality rather than real creativity.

- ACTIVITY; desire for; creative/ ARTISTIC; aptitude – Chir-f
- CHEERFULNESS; dancing, laughing, singing, with/ Music, from– Chir-f
- DANCE, desires to– Chir-f, Cor-c
- DELUSIONS, imaginations; angels, of/ Flowers, pink granite, made of/ Jewelry, of/ Landscapes; beautiful, of/ Things, objects; golden – Stoi-k
- DREAMS; colorful; orange/ Yellow– Chir-f
- DREAMS; music/ Singing, of – Stoi-k
- SENSITIVE, oversensitive; colors, to– Stoi-k
- SING, desires to– Chir-f

Expression words: *Extremely sensitive, Startling from Noise, Sensitive to pain – suffering – injury, Sensitive to insult and vexation, Amazed by feeling, Amazed by touch, Amazed by sound, Amazed by beauty and colors, Ticklish, Sensual, Staggering, Astonishing, Overwhelming, Stunning, Rush of adrenaline, Surprised, Pleasurable, Touch everything, Beauty, Color, Artistic aptitude, Like dance, Like singing, Like music, Like physical rhythmic activity*

(2) Exaggerated senses or Active sixth sense

Like the sponges, cnidarians also have heightened senses. They can sense more than normal beings can, so it looks like they have an active sixth sense. They can sense dimension, shape, color and sound. Mainly they feel their body and body parts. They have a feeling of electric impulses.

- DELUSIONS, imaginations; body, body parts; smaller/curtain hangs between her and others, a heavy/ Dissolving, she is/ Divided; two parts, into/ Double; he is – Stoi-k
- DELUSIONS, imaginations; Voices; hears; night/ Whispering; him anything, someone– Chir-f
- DREAMS; outlines, black and white– Chir-f
- DREAMS; tongue; large, too– Chir-f

(3) Irritable, Angry, Rude, Abrupt
Cnidarians are the first animals to have a decentralized nerve net and are able to sense and react, but they do not have a brain and so are not able to store their memories. In patients this expresses as a feeling of being ready to react at any moment. They are very impulsive, but as the situation passes away they cool down very easily and they do not keep anything in their heart. They may get angry with anybody but the very next moment they forget everything are ready to talk with the perceived offender. Cnidarians are armed with cnidocytes, so they are always ready to sting. The cnidarian patient is also always ready to lose their head. These qualities color the cnidarians with irritability, abruptness and rudeness.
- ABRUPT– Chir-f, Stoi-k
- ABUSIVE, insulting– Cor-r (ABUSE agg., ailments from/ DREAMS; criticized, being – Stoi-k (DREAMS; abused; being– Chir-f)
- ANGER; trifles, about/ Easily/ Sudden– Chir-f (ANGER; violent/ RAGE, fury; striking, with – Stoi-k)
- CENSORIOUS, critical; friends, with dearest– Chir-f
- CONTRADICT, disposition to– Chir-f, Medus (CONTRADICTION; intolerant of– Stoi-k)
- CURSING, swearing, desires; amel/ Pains, at – Cor-r
- DREAMS; insults/ Mortification, humiliation – Chir-f
- IRRITABILITY; anxiety; with/ Pain, during/ Questioned, when– Chir-f
- IRRITABILITY; trifles, about– Chir-f, Medus
- LAMENTING, bemoaning, wailing; pain, from– Cor-r
- QUARRELSOMENESS, scolding; pains; during– Cor-r
- SENSITIVE, oversensitive; rudeness, to– Stoi-k

Cnidarians have neurotoxins in their stings that paralyze their prey. Sponges and echinoderms do not have toxins (with some exceptions!!), so their toxic quality expresses in patient as extreme aggression, jealousy, cruelty, hatred, feeling of being attacked, feeling of being perused etc.
- CRUELTY, brutality, inhumanity– Chir-f
- DELUSIONS, imaginations; poisoned; has been, he– Cor-r
- DICTATORIAL– Chir-f, Stoi-k

- DREAMS; plundering– Stoi-k
- FEAR; attacked, of being– Stoi-k
- FIGHT, wants to– Chir-f, Stoi-k
- HATRED; persons, of; enjoying life, who are– Medus (JEALOUSY; people around, of– Chir-f)

Expression words: *Always ready to lose your head, Irritable, Angry, Impulsive, Abrupt, Ready to react at any moment, Triggered out easily, Soon angry and soon cool, Extreme aggression, Jealousy, Cruelty, Hatred, Feeling of being attacked, Feeling of being perused, Numb, Paralyzed, Tingling*

Reaction – Withdrawing
Cniderians' first reaction in the presence of danger is to withdraw. Withdrawal is more pronounced in Anthozoa and Hydrozoa, but is present to a lesser degree in the Scyphozoa and Cubozoa. The Anthozoa and Hydrozoa are less evolved, having some similarities with sponges. The feeling of withdrawing expresses in patients as curling up, becoming smaller, contracting, paralyzed, stupefied, becoming numb and becoming insensitive etc.
- ABSORBED, buried in thought– Stoi-k
- CONFUSION of mind; alcohol, alcoholic drinks agg/ Beer, after– Cor-r
- CONFUSION of mind; talking; amel/ Intoxicated feeling – Stoi-k (Cor-r)
- CONTENTED; oneself, with/ PESSIMIST – Stoi-k
- HURRY, haste; alternating with; withdrawal– Stoi-k
- STUPEFACTION, as if intoxicated; wine, after– Cor-r

Reaction – Counter Attack- Prickling, Poking, Shooting, Darting
Cnidarians have cnidocytes that shoot out small harpoon-like barbs and deliver toxins. This expresses in patients as burning, prickling, smarting, pocking physical sensations.

Motion and Movement [Flying, Floating, Rhythms, Balancing]
Cnidarians are the first animals able to freely swim as medusa, so movement is very characteristic of this group of remedies. This issue expresses in patients as a sensation of flying, floating, gravity-less, free moving, air-like, hovering, movement in slow motion etc. They like walking, wandering, and any rhythmic movement. Their main focus in movement is to keep balanced, so they like motorbike riding, bylining, etc. They prefer dancing, especially that which has rhythms and balancing of the body. Movement is more pronounced in Scyphozoa, Cubozoa and Hydrozoa then non-medusial Anthozoa.
- CHEERFULNESS; dancing, laughing, singing, with/ Music, from– Chir-f
- DANCE, desires to– Chir-f
- DELUSIONS, imaginations; floating; air, in/ Light, incorporeal, immaterial, he is– Chir-f

- DELUSIONS, imaginations; motion; things happen in slow motion– Stoi-k
- DREAMS; flying– Chir-f
- DREAMS; wandering– Chir-f
- WALK, walking; amel– Cor-r

Symmetry

They are first animals to have body symmetry, so they have a fixed and regular shaped body. This expresses in patients as a like of detailing. Their work and personality have some order. They like orderliness, so they appear fastidious. They like work in which you have to apply analytical mind. They are more systematic than sponges.
- CLEAN, desire to– Chir-f
- FASTIDIOUS– Chir-f
- OBSTINATE, headstrong; duties, in performing irksome– Chir-f
- ORDER, desire for– Chir-f
- REST; cannot, when things are not in proper place– Chir-f

Other authors have also contributed to our understanding of this phylum. Jo Evans also adds the following characteristics of the cnidarians[98]:
- Ultra sensitive
- Extreme sensitivity to pain
- Weakening of boundaries and inhibitions, as if drunk
- Nervous exhaustion
- Out of body sensations
- Lack of empathy
- Clairvoyance
- Strong right brain dominance (more creative and emotional)
- Numbness, detachment slowness
- Lack of tolerance and adaptability
- Completely connected or completely cut off
- Sensation of dismemberment of decapitation
- Violence, beaten, broken, torn apart sensations

She also describes some frequently observed sensations of the cnidarians[99]:
- Burning
- Itching
- Stinging
- Crawling on the skin
- Pins and needles
- Electricity
- Tingling
- Numbness
- Throbbing/pulsing

- Feeling like jelly. Weakness, collapse
- Emptiness, hollowness
- Dislocation, dismemberment, disorientation
- Swelling, enlargement, elongation
- Squeezed, pinched, compressed, constricted
- Pulled or drawn backward, pressed downwards
- A sensation of a foreign body inside

I would add in:
- Fatigue
- Poor memory
- Confusion
- Difficulty with organization

The sensations and characteristics describe above will serve as further confirmation of the remedy group, in patients displaying the core problems of this layer.

For being one of the less prominent phyla, the cnidarians are actually quite well represented within homeopathy. Most subphyla are represented, with the exception of the Staurozoa. The following table contains a list of the proved and unproved remedies within Cnidaria.

Subphyla	Proved Remedies	Unproved remedies
Anthozoa	Anthopleura xanthogrammaica[G] (antho-p), Corallium rubrum[G] (cor-r, Gorgonio nobilis), Corallium nigrum[H] (cor-n), Diploria clivosa[E] (diplo-c), Fossil Fungia Coral[A], Fossil Dimorphophaestra[A], Heteractis malu[R,F] (hetera-ma), Stichodactyla gigantea (stich-g)[H,F]	Calliactis parasitica[R] (callis-p), Eunicella singularis[R] (euni-s)
Hydrozoa	Physalia physalis [R,F,H] (physala-p)	Hydra oligactis[R] (hydra-o)
Scyphozoa	Medusa[G] (medus, Aurelia aurita)	Chrysaora quinquecirrha (chrysao-q)[H]
Cubozoa	Chironex fleckerii[H,F,R] (chir-f)	None
Staurozoa	None	None
A-Available at Ainsworth's R - Available at Remedia F – Available at Freeman's G - Available at most pharmacies H - Available at Helios E – Available at Enzian		

Proving Suggestions:

Unusually for one of the less prominent phyla, most cnidarian subphyla have at least one proven representative. The exception is the recently delineated Staurozoa subphyla which so far lacks any proved or unproved remedies.

However, the Cubozoa, Scyphozoa and Hydrozoa sofar have each only single representatives within the materia medica. The anthozoa as well have only a few species proved. Only two species of coral have any information whatsoever relating to their homeopathic uses.

As such provings of all cnidarian remedies are strongly recommended, with a particular emphasis any of the hydrozoa, scyphozoa, and cubozoa as well as any corals. Staurozoa are also strongly encouraged, due to a complete lack of these organisms within the material medica.

Some particularily interesting cnidarians to prove[100] would be parasitic species, such as *Polypodium hydriforme* the only known Cnidarian intracellular parasite, *Actinoscyphia Aurelia*, an anemone which actively engulfs prey and releases bioluminesct jelly, *Edwardsiella lineata* a jellyfish with parasitic larvae which can implant in humans and cause an itching called sea bather's eruption, or *Cassiopeia*, a photosynthetic jellyfish which spends most of its life upside down, exposing the commensal algae in their tentacles to the sunlight. Another potentially interesting proving would be *Turritopsis nutricula*, an "immortal" hydroid hydrozoa which does not age, but instead reverts to its larval phase, rematures, and begins life again, regenerated.

Anthopleura xanthogrammica (antho-p)[101]:

Anthopleura xanthogrammica, also known as the giant green sea anemone is a small species of anemone native to the West coast of North America from Alaska to Panama. It lives in lower intertidal zones and feeds on small crustaceans and other animals that come within reach of its tentacles. It can move, very slowly if it needs to relocate. This remedy was proven by Cynthia Sheppard in a blinded proving without placebo in 2005[102]. Most of the information in this monograph derives from this proving, with some input from Jo Evans, who produced a wonderful summary of this remedy in her book "The Sea Remedies"[103]. A trituration proving was also done by Roland Geunther[104]

Key Issues: The main issue in this remedy is <u>connection with others</u>. This state of connection is extremely desirable to the being representing the divine space before layer 1 from which they came. <u>Anything that disrupts or offends others and makes them withdraw from the being will disrupt the integrity of their selves , which are still dependent on the feeling of being united</u>. Hence issues of social functioning as well as taboos such as homosexuality, pornography, cross-dressing, dirtiness, disease, deformity and so forth become prominent. These issues aren't the core of the remedy however. They represent issues that would result in <u>social ostracizing</u>, and that <u>detract from their sense of connection to those around them</u>. Other issues, that are taboo or disapproved of in a given social context are likely also to come up, depending on the individual patient's environment. This is especially clearly elucidated in the following extract from Shepard's proving:

> First thing in the morning, while still lying in bed, I was very aware of the "in between" place, the place between sleep and being awake. I was totally aware. I could slowly open one eye briefly, I could hear, I was aware of the warmth of the bedding and of the clothes on my skin, and I felt safe enough to pace the panic that I often feel during the day. I realize that the panic is the result of being afraid I won't meet the demands put on me. A big part of the panic is fear of doing something that would result in causing suffering to others or to myself. When I face what the suffering means by asking, "How does the suffering feel?" The answer was that suffering meant the breaking of connections with people and other living forms around me. Further, the breaking of these connections felt like being not connected not only to my family and all living forms such as the cats, the trees outside, the squirrels, raccoons, etc., but also to be separate to their higher consciousness. That is when I realized that the "I" does not exist! The "I" is the immense connection of networks with all living beings around me to the living force. Without that connection there is no living force and no "I". At that moment there was no panic, there was just that moment; the "I" had gone. Now, that I am in the awake state and I am writing this, there is the "I" and there is the panic, but it is less. I now think, "What was that?" It was not death. It was just consciousness. It was the realization that the "in between" place is not dreaming, it's like a deep meditative state. There is no death. I'm finding it hard to recreate in writing that feeling sensation that happened in the "in between" place. The awareness developed into realizing that there was no darkness anymore but a bright shiny overwhelming bright light instead of the dark. I was creating the dark by shutting out the light. There is still an awful lot of light in me now, though it is fading[105]

Mind: Forgetfulness, poor organization and planning, Making mistakes in writing and speaking. Lack of concern with timing. Carelessness, slow thinking, poor mental function. Lightheadedness. Spacy feelings, feelings of unreality. Numbness, no emotion, not able to access feelings. Feeling disconnected from others. Blocked Intuition. Irritation with others , Aggression, in response to others' transgressing boundaries. Irritation in response to disconnection. Alone and abandoned (in reponse to disconnection). Sensitive to noise and sensory stimuli. Social structure very important. Social structure completely unimportant, disdainful of rules. Mind functioning like a reptile, no social consciousness. Feelings of heaviness.

Main Physical Issues: Exhaustion, heaviness of body, fatigue, slowness. <3pm, >by the ocean, <noise, <heat. Heat with internal chill. Tissue affinities with CNS, muscles, (especially back symptoms), skin, gastrointestinal tract, digestive system, urinary tract

Clinical[106]: Sunstroke, post viral fatigue, heart issues, sciatica, diarrhea, dehydration, sensitivity to noise and sound, chronic fatigue, nervous exhaustion, menstrual and ovarian issues

Miasm: Tubercular, Sycotic

Color preference: ?

Remedy Comparison: Other cnidarians. Layer 2 remedies in general.

Corallium nigrum (Cor-n, Antipathariae sp.)[107]:

This remedy was made of a specimen from the genus Antipathariae[108]. The exact species is not known. Despite the name of this remedy, this branch of the anthozoa is actually more closely related to the anenomes, rather than the corals. Black coral is the state gem of Hawaii. It is often used for jewelry and gifts. The remedy has not been proved, but one of my own cases responded well to this remedy, allowing the presentation of a preliminary picture.

Key Issues: The key issue of this remedy is the social environment. The being feels very strongly overwhelmed by social conflict and by others drawing upon their resources and imposing upon them socially, destroying their sense of self. They feel they have limited resources available to care for themselves, and that others requiring things from them drains the energy they need to preserve their sense of self. Their response is to withdraw into a safe space and isolate themselves. This remedy is very similar to Heteractis malu but seems to be more sensorial in its orientation.

Mind: Fearful of conflict, marital and relationship problems. Problems trusting.

Main Physical Issues: Fatigue

Clinical: Lyme disease and chronic fatigue syndromes.

Miasm: ?

Color preference: ?

Remedy Comparison: Corallium rubrum, Fossil Dimorphastrea, Heteractis malu, other cnidarians, Silica, Pulsatilla

Case:

The patient is a woman in her 20s who presents as a referral from another ND via Skype for casetaking. Her main presenting issues were fatigue, trust issues and lack of balance in her life.

The fatigue literally felt like she was being drained. She noticed it specifically at her work. She described it as a feeling of heaviness, and then she described her reaction to this heaviness as checking out of her life and not knowing where she went. She just wanted to sleep but found sleep unrefreshing.

At this point we shifted to discussing her trust issues. She described some specific trust issues she had with her husband, then described how she saw trust. She described her life as the building of a fortress around herself. Trust was allowing someone inside the fortress, or allowing herself to come outside.

When asked why it was so difficult to come outside of the fortress, she described it as scary, not safe and an unknown territory. She felt like she did not understand anything, and that no one taught her how to deal with the world outside her fort. This left her in a dangerous situation.

When asked to describe the feeling of that dangerous situation, she described it as a shock, a numbness and a "checking out". She felt tingly, sick to her stomach and hot. It reminded her of a niacin flush.

She then went into the source. She saw a place in the dark where her brain shut off, she couldn't think, and she was frozen and couldn't move. She felt that she was inside a small space, only slightly larger than her body. This space was dark and quiet, but safe. Her brain shut off, and emotions stopped working. She experienced an image of herself inside a cement room, with cracks in the walls. Here nobody could find her and hurt her.

The opposite of this sensation was a feeling of a wide-open space. It was so bright that it hurt her eyes. Not knowing what to do there, she reacted in two ways.
- (1) To crouch down and become hyperaware of anything, looking for danger
- (2) To become happy, and free, twirl around with joy

This happy and free feeling was described as being light, colorful, childlike, playful and magical with no cares in the world,

I had a very strong suspicion about her remedy and so asked her to describe her sense of self. She described herself as flip-flopping between a sense of self and no sense of self. She felt that when she was in the place of being free and happy, experiencing the openness, she had a sense of self. She lost her self, however, by getting involved in other people's business, and then would check out of life by retreating into her dark space.

This final sense, I think, is the emphasis of her case. She displayed some very clear source information relating a cnidarian sense of unprotectedness in response to external stimuli, and she reacted to it by withdrawing into a tiny little dark closed area. This clearly pointed to a coral. While the memory symptoms may have suggested a brain coral, the lack of symptom correspondence between this case and the Diploria trituration led me to look elsewhere. The lack of correspondence between this case and my unfolding trituration of Corallium rubrum also led me to look at other remedies.

I settled on Corallium nigrum. I prescribed her 1m in liquid 1 tsp daily, taken succussed.

Followup 1:

The patient and I met via Skype in late December 2014. The patient reported a distinct emotional shift since beginning the remedy. She was more conscious of her tendency to isolate herself, and she had a very bad cold (there was quite a terrible cold circulating in most of Canada at this time).

However, the patient now had a few instances of feeling completely free of anxiety in social situations. She had instances of feeling carefree, as if she had a <u>bubble surrounding her and protecting her</u>. She described this bubble as invisible, allowing her to have fun and do what she pleased without worrying what others thought of her.

I asked her about the pervious state she experienced, and she described it as considerably lessened. However, she was beginning to consciously deal with several problems in her relationship with her husband. She was still unsure of what she wanted to do, and felt very confused about her feelings.

We can see the patient's state considerably improved after administration of Corallium nigrum. Particularly interesting is the description of a bubble that protects her from social influences. The emergence of her marital conflict into the realm where she perceived action was possible is also a very good sign. She is now strong enough to begin to address the problems in her relationship. Her sense of self is now conscious enough to deal with fundamental conflicts in her family unit, a very encouraging sign. She had not yet completely overcome her state, but had noted no plateau of effects from her remedy yet, indicating that the current potency was still effective.

I reassured the patient that this process was healthy, and I told her to have faith in herself and her own ability to sort through this relationship problem. I told her to continue the Corallium nigrum 1m, 1 tsp succused per day. I also told her to avoid deciding on her relationship status until she was completely sure about what she wanted to do. I also prescribed 30 mins of yoga nidra per day[d].

Followup 2:

We followed up again via Skype in January 2015. The patient reported that she had regressed since the last visit. At this time, instead of a feeling of being overwhelmed, she reported a feeling of profound irritation. She felt extremely irritated whenever people drew upon her, feeling she had nothing else to give to them.

She also disclosed that she had been diagnosed with a low grade squamous intraepithelial lesion on a recent pap, revealing a history of chlamydia and HSV-1 infection. She went deeper into the circumstances of her and her husband's marital issues for the first time as well.

Her move from being overwhelmed to being irritated when others draw upon her resources indicates an increase in vitality and health. Irritation is a far more energetically demanding reaction than simple overwhelm. This indicates that she has far more vitality available. I suspected as well that her LSIL may have been related to her previous history of STIs.

We increased the potency of Corallium nigrum to 10m, 1 tsp taken in a liquid dose, succused daily. I also put her on pascoefem, and began a course of CEASE therapy for her previous chlamydial infection.

---

[d] Yoga nidra is a meditative technique based on body consciousness.

Corallium rubrum (Cor-r, Gorgonio nobilis)[109]:

Corallium rubrum is a small species of coral found mainly in the Mediterranean. It is a member of the Alcyonaria subclass of the Anthozoa. It was commonly used for jewelry and medicine in the Greco-Roman world, and this use continued up to the 19th century. It was used for seizures, profuse bleeding, gonorrheal diseases, kidney stones, and dental caries[110]. The remedy was proved by Attomyr in Germany before Hahnemann's death[111] and appears in Allen's encyclopedia. The mental/emotional picture, however, has not yet been adequately elucidated. As such I selected this remedy for a trituration. Most of the mental picture derives from my own trituration, given at the end of the chapter. Most of the physical picture was summarized from Murphy's Nature's Materia Medica[112]. The C1-4 of my trituration is presented below this monograph.

Key Issues: This remedy has a heavy sensory emphasis. The being intensely feels its sensory input, and lacks the strength of self needed to withstand its perceptions. The <u>sense of self of the being dissolves due to the intense sensory perception</u>. The being responds by <u>withdrawing into a small safe and quiet environment</u>.

Mind: Irritability, intoxication, as if drunk, fear of suffering. Abusive and insulting. Note that the common rubric ascribed to Cor-r "Delusion, feels as if newly born in the world and overwhelmed with wonder at the novelty of his surroundings[113]" is in fact incorrect. This rubric is properly applied to Cori-r, Coriaria ruscifolia[114]. This rubric, however, is quite intriguingly serendipitous to the themes of the Cnidarians.

Main Physical Issues: Respiratory issues. Dry spasmodic cough. Cough worse eating. Vomiting of stingy mucous after cough. Cough with blood. Rapid coughs. Violent thirst, Craving salt and acid. Congested or tearing headache, worse stooping and cold air. Sexual weakness, wet dreams, white coat on tongue during whooping cough. Nasal catarrh, profuse mucous. Nausea with dry tongue.

Clinical: Whooping cough. Coughs. Catarrh.

Miasm: Tubercular

Color preference: 8C

Remedy Comparisons: Corallium nigrum, Cnidarians, Spongia, Cocc-c, Drosera, other cnidarians

Corallium rubrum Trituration: Unification of the Self

This trituration was made from a sample of Corallium rubrum purchased from a small online jewelry store in Spain.

C1-3:

In these three initial triturations a clear picture emerges of a being with a very weak self. This self is buffeted by sensory impressions it cannot tolerate. Light is too intense, sound too powerful, etc. These impressions are too strong for the beings self to deal with, and result in the being feeling overwhelmed and the integrity of it's self being destroyed. This is described as a feeling of being jittery, quivery, fizzling, dissolving (DDX actinides) and falling to bits. The being's response to this is to withdraw, specifically to withdraw into a small cave-like enclosure in which they feel safe and secure. A place where they can finally relax.

C1:
Preformed November 19th 2014:

- I feel strange. I'm quivery inside and weak
  - A very dizzy feeling
- I feel weak, shy, incapable, passive
- I feel a lot of activity in my digestive organs
- I'm very uncertain. I feel quite afraid of my surroundings
- I'm afraid. I don't really know why, but I am afraid right now!

C2:
Preformed November 22nd 2014:
- I feel uneasy with my surroundings. Perception is intense
- Light seems intense
  - I feel a flushing of heat on my cheeks
  - I feel jittery and quivery inside
- I feel a sensation of heat inside my chest and on my skin
- It is this internal jitteriness that makes me anxious. I'm afraid. I'm afraid of everything
- I feel like I have to keep my guard up to guard myself against whatever is making me jittery. It is a very rigid feeling
- I feel weak and vulnerable, but vulnerable to what?
  - The external world
  - I feel like withdrawing
- I feel hot all over, there are heat flushes on my skin
  - Intense nervousness
- The heat is starting to lessen and release

C3:
Preformed November 23rd 2014:
- The fear is more intense now. I'm afraid of my surroundings
  - I feel a quivery weak sensation inside.... Which is surrounded by a shell
- I feel very weak inside- but I hide myself in a shell to avoid this weak feeling
  - I hide myself in this shell and feel that I am less exposed
- I feel vulnerable. Life is intense! I want to withdraw into myself, into my own little cave to escape it. Sensation is just too intense
- I feel very mentally withdrawn and hesitant
- Perception of movement, noise. It all seems louder to me. I cringe backward and withdraw from it. I just want quiet and peace
  - Quiet is the lack of sensory stimulation. It feels dark, quiet and peaceful. I can relax there
- I just feel like the world is too intense right now to connect or engage with. I don't feel strong enough to stand it all
  - The world will overwhelm me!
- My body feels soft and vulnerable
  - I'm not able to deal with life. It is too much input

- My self can't deal with it all and stay intact!
- I will fizzle and dissolve!
- I will fall into bits!

C4

This level of the trituration describes the sense of jitteriness in greater detail. The forces holding the individual components of the being together are not as strong as they need to be to withstand the full force of sensory impressions. A metaphor of evaporation is used to describe the impact of sensation. It is as if sensation imparts an energy to components of the self, and the self is not bound together strongly enough to maintain its wholeness. This remedy strengthens the forces holding the being together, allowing them to withstand sensation. Sensation is still too intense and the being still feels a need to withdraw from it, but this is not due to the dissolution of self. Instead it is due to the perceived intensity of the stimulus.

Preformed November 24th 2014:
- There is a sudden feeling of heaviness around my head
- Life is intense. I feel sensation is very strange
  - I feel odd, out of sorts
- Then suddenly, the intensity diminishes, first on my right side
  - The sensation on my right is still intense, but I feel like I can deal with it now
- Now the left side is following. Sensation isn't so disturbing there
- Sensory input is still very intense, but I feel like it no longer grates upon me. It's just a lot!
  - I don't feel threatened by it. The anxiety is lowering!
- I feel good. I no longer feel jittery inside!
- It is as if the external world no longer threatens me, and causes my insides to react with fear
  - To jitter
- I still feel like I want to back off from loud noise, but it is not affecting me deeply. It's just too loud!
- That jitteriness. It was like static, or like a flickering. It was like my insides were moving relative to one another.
  - They shouldn't be doing that!
- If it gets worse it will be like evaporation. The particles making me up will move so fast they will break away from the whole structure. It's like whatever is holding me together isn't all that strong
- I feel really good right now!

C5

This level of the trituration returns to the jittery feelings of C1-3. The world is again perceived as being too intense for the being to handle but at this level a new reaction is present. In addition to withdrawal, the possibility of a hostile or aggressive defense against the world is presented. The being feels irritated to and hostile toward the world. This was accompanied by a source sensation of having spikes with stingers, which can attack anything that gets too close.

Preformed November 26th 2014:
- I'm feeling a little jittery, specifically in my abdomen
- I'm feeling nervous and anxious again
- I'm irritated, irritated with the world outside of myself
    - The world is so loud and aggravating!
- I'm angry. How dare the outside world disrupt my sense of self!
    - All of this input. It's too much!!!!
- Life irritates me. I'm reacting with anger to outside stimuli
- It's like I am reacting against the external world
    - I feel this area of defense around my body like this series of spikes ready to defend me
        - The spikes feel like they have stingers! Stingers to attack anything that comes to close!
- I feel like I have this hostile reaction to the outside. I cannot handle that much of it so I am hostile to it, or I withdraw to keep myself safe
    - Hostility is, after all, a defense

Diploria clivosa (diplo-c)[115]:

Known colloquially as the knobby brain coral, Diploria is a colonial coral species native to the West Atlantic and Caribbean. It forms domes on the ocean floor upto 4 feet high. They use both photosynthesis and tentacles to obtain food. Growth is slow, and big colonies can be over a century old. A sample of Diploria was triturated by Anne Schadde in 2003[116]. I have again relied upon Google translate to translate the information on Schadde's website. I cannot guarantee the complete accuracy of Google's translation. I eagerly await a human translation. I have included the full text of the translated trituration at the end of this monograph, edited for English grammar.

Key Issues: The key issue of this remedy from the trituration data available seems to be symbiosis. The being feels a great symbiotic relationship with those around them, stabilizing themselves. This relationship is disrupted by chaos. This chaos results from willfulness, from going against higher will with one's own will. This chaos disrupts the harmonious relationship needed for symbiosis and hence disrupts the integrity of the self.

Mind: Poor mental functioning, confusion, difficult thinking, poor memory. Timelessness. Weak-willed or strong willfulness. Going against higher will with chaos resulting.

Main Physical Issues: Dryness of body parts. Dry mouth, difficulty swallowing, burning eyes. Dry cough. Headache left temple, > sleep.

Clinical: Based on existing data, this remedy may be useful for dry coughs.

Miasm: ?

Color preference: ?

Remedy Comparisons: Bryonia, other Cnidarians.

Brain Coral from the Dominican Republic
By Anne Schadde[117]

Name: Diploria CLIVOSA
The substance:
1. Reeking of ammonia, sulfur
2. Amazingly hard
3. As if worms in the small channels
4. Under water they looked larger

Associations:
From the skulls, the worms grow
It smells channel
That in the brain by channel smell ??
In prep course the brain smells far less than other parts

Trituration C1
Ride on the dive boat
It was a little careless boat with the instructor at the dive sites
into the open sea, anchored somewhere
Anticipation; can dive
A new dimension of what we see
How can I find himself
Underwater, the self-perception is another
Beautiful make underwater headstand
I looked at these coral upside-down
Total fascination
Looks like brain from Prep Course
Similar to the moment Steiner discovered the mistletoe[e]
The colors under water was larger and more pink, more match actual brain

---

[e] I believe this refers to the discover of Mistletoe as a Cancer remedy by Rudolf Steiner.

I start to hyperventilate
It's getting warm
The fascination of brain coral is that it could be a medicine that could act on the brain
Spontaneously, I thought, Alzheimer's and MS ....
(Because of the herd, where the worms crawl)

Corals are anything but dead, cities are for the fish
It evokes complex meander structures
Fold
Outside and inner it has a highly complex structure

What are corals?
Have stinging cells and therefore belong to the group of cnidarians
(jellyfish, polyps, sea anemones and sea fans, the flowers animals)
Then soft corals are often distinguished from stony corals

Stony corals powerful calcareous skeleton
On the skeletons of the other sea creatures live

Symbiotic organism ...

Very sensitive to stress
The landlord throws his subtenant out during stress

SYMBIOSIS is the core concept: external stimuli, internal control .....

Coral bleaching: the coral loses its color when the stress is applied too long for the animals
If the stress is too large, he throws the helper[f], without which he can not live out.

What is flora? The flora is given by the mother.
In Germany flora is a taboo subject. We consist of more than symbiotic body cells.
Because we are under stress we throw first with the AB[g] lodger out ...
The fact that we throw out all the symbionts, we are too clean.
Environment shifts – the organism is changed

Pictures of all sorts of different types of fish that hide together even if they do not swim side by side. The coral and a generous hosts.
All kinds of corals stand together like a forest

---

[f] I think this means the algae upon which the coral depend for nutrition
[g] I suspect this means we use antibiotics too often and throw out our commensal flora

But the brain corals are always isolated
No matter what stage of growth: a little girl is young and big old

The coral is the home for fish, such as ferns for the animals in the woods
The instructor says, if you see sharks, dive into the coral!
Sand is actually shattered shells
Mussels are passed through the coral ... be transported as coral
Sand is grated rock
The algae are the water filter

Physically:
Water, thirst, dry lips, rear pharyngeal wall totally dry
Throat sticks when swallowing
Feeling of a lump in the throat
Elation, the same feeling as on the boat now or in the pure driving (?)
Right nostril ....
Neck is still dry .....
Dear water instead of juice .... why not champagne?
Fair: all possible associations: Word twist
It is a word and there are cabaret changes
As if to drink champagne
Laughter

Totally confused, arithmetic does not work
Where have you left your brain ??
It's harebrained ...

Trituration C 2

Burning eyes
joking about word analogies
Brain coral looks soft under water
Confused with Amber Room "Coral Room"
The coral is female, she is round;
Time does not matter
Thoughts meander to the structure of the surface
Complacency, not really want to do any theorizing.
Again and again thoughts go to order and chaos
By rubbing[h] time is disoriented
Topic: Electric Conduction - and in the water?
Right nostril runs
biting teeth
Coral as host. Symbiosis, mother-child problem

---

[h] Scraping within the trituration, I assume

Reluctance to speak, but in serenity
loses when triturating the spiritual thread
teasing comments
Mood intensifies the emotional intelligence
you know anyway what is being said
"Everything under one roof"
Thirst, great gulps, salty taste in the mouth, dry mouth
Dry mouth, back, swallowing goes no further.
Listen to words and do not understand it, but trying hard
Loss of connection;
Increasing distortions word
Blurred words
What is coral when she dies
Cough; as if dryness shares would slipped deeper into the larynx
Memory for spoken word is weak

Trituration C 3

Topic: Symbiosis
"Pained face"
is taciturn: I only listen to ....
Round - we do not have in hand
Structure will
The will comes from a higher power
We are forced to it's will
It is difficult to find the order
We spend a lot of time to brace against it and not to accept the order[i]
It stinks again ....

Follow-up

I cannot concentrate.
I am somehow mentally tired and quite beside me.
I'm slow,
Have a weird head pressure, it is not a headache, rather dazed.
But I slept soundly and well.
When I went home, I almost lost my way.

Still cannot quite focus.
On the return trip headache left temple,> Sleep

---

[i] I think this phrasing refers to the prover feeling that there is a higher will and an order associated with it and that a great deal of energy is expended trying to resist this will.

Fossil Dimorphastrea

Dimorphastrea was a coral species that lived from 189.6 to 5.6 million years ago[118]. It was a scleratinan coral, and was proved by Martine Mercy[119]. I have also included some clinical information from Mercy's book on toxicity[120].

Key Issues: The main issue in this remedy seems to be the <u>environment</u>. The being <u>is extremely vulnerable to outside environment, with few barriers</u> to prevent it from <u>absorbing all kinds of toxicity.</u> This toxicity gives it a feeling of <u>intoxication</u>. This toxicity causes the being to <u>withdraw into a space of isolation and seclusion</u>. However the being becomes <u>extremely lonely</u>, desiring the company, consolation, communication and exchange of an active life.

Mind: Takes on invasive stress from people they are close to. Feel their space to be easily invaded or overcrowded by the emotions and issues of others. Feels as if intoxicated. Averse to others around them. <Intrusiveness, <speaking of others, <presences of strangers, <rivalry, <competition, <limited personal space. Delusions of being beautiful, delusions things are beautiful. Loves communication, but in a silent environment without interaction. Feels alone within a group. Better for independence. Desires liberty. Worse limited space. Loneliness. Feels she doesn't belong, feels neglected. Better for consolation. Lacks confidence, Indecision. Fastidiousness. Cannot work when things are chaotic. Cannot rest when things out of place. Quiet. Longs for tranquility. Dislikes change and noise. Easily impressionable. Confusion of identity.

Main Physical Issues: Environmental toxicity, chemical, heavy metal. Easily affected by external toxicity, such as EMFs, pesticides, allopathic medications, alcohol, cosmetics and vaccines. Lack of sexual desire in women. Vaginal dryness. Painful intercourse. Insomnia when mind is overburdened. Wakes at 2 am (liver time). Dysfunction in right cervical spine, occiput and right ear. Eczema. Poor digestion, bloating <Pressure, <nuts, <chocolate. Great thirst. Heaviness/numbness in left arm. Poor circulation in legs. Cramps in calves. Shoulder pain.

Clinical: Environmental toxicity, particularly reduction of absorption of chemical toxins. General drainage of toxins. Drainage of organic chemical toxicity. Side effects of allopathic medicine, especially statins. Alcohol toxicity. Vaccination toxicity. Toxicity from Cosmetics.

Miasm: ?

Color Preference: ?

Remedy Comparisons: Other Anthozoa. Other cnidarians. Other fossil remedies. Agaricus. Corallium nigrum. Echynocorys. Heteractis malu. Complimented by Calamites. Antidoted by Lapis Immersion.

Fossil Fungia Coral[121]

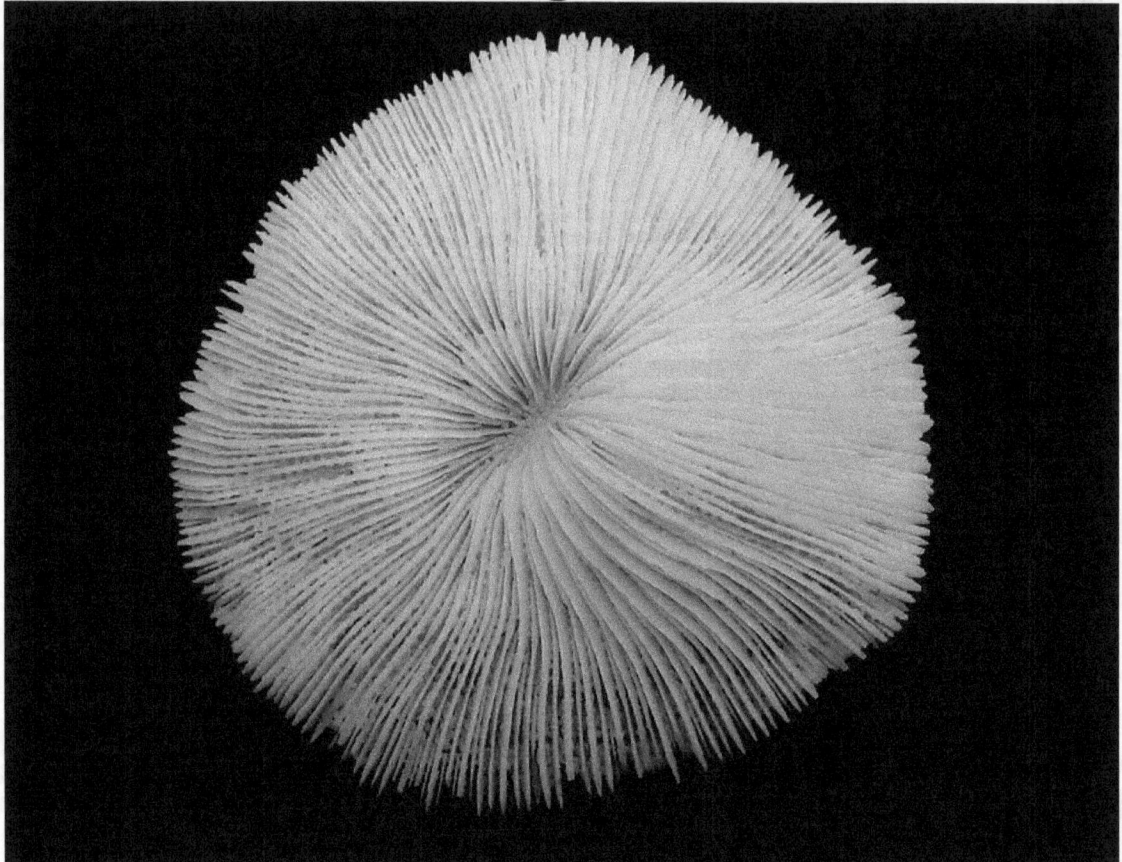

Fungia is a genus of extant corals native to the Indian and Pacific Oceans. They are scleratinan corals and are popular in the aquarium trade. Fungia are generally solitary, but one species, *Fungia simplex*, is colonial. Some species also are capable of changing sex. The remedy was proved by Martine Mercy form a Fungia fossil she obtained using both a painting proving methodology and a traditional Hahnemannian proving[122] [123].

Key Issues: The key issue in this remedy seems to be one's ancestral inheritance. The being has no barriers between itself and its ancestral inheritances and cannot move beyond them into its own life. The being floats above its day-to-day life, not fully connected to any part of its life due to its immersion in the issue of its ancestry. As such it cannot completely fulfill its own roles in life, notably motherhood (though I also suspect that other roles may be unfulfilled as well). Indeed, it barely even feels it's own life.

Mind: Indifference. Will not get involved in arguments. Never in deep or profound turmoil over anything. Dominated by ancestral and cultural patterns. Limited by the past. Patient is incomplete, cannot completely fulfill their life. Failed at motherhood. Had to work. Could not care for children or breastfeed.

Main Physical Issues: Strong effect on the mucous membranes. Drainage of the external ear (followed by emerald or Hep Sulph). Headaches derived from toxicity. Yellowish green profuse mucous. Sensitivity of throat to the atmosphere. Catarrh. Nostril Ulceration. Laryngitis with green mucous. Affinity with Gallbladder Channel.

Clinical: Drainage of mucous membranes. Drainage of the respiratory tract

Miasm: ?

Color Preference: ?

Remedy Comparissons: Other Anthozoa, Other Cnidarians. Other Fossil Remedies. Emerald. Hepar Sulph. Sulphur. Pyrogenium. Agaricus.

Cases: Two cases are reported in Mercy's 2009 book. One was a female student with muscular pain and digestive issues overwhelmed by her classes. The second was a woman with a number of relationship issues.

Heteractis malu (hetera-ma)[124]:

This anemone, also known as the delicate sea anemone, is native to the Pacific, being found in Hawaii, Japan and Australia. It is a member of the Stictodactylidae along with *Styctodactyla gigantea*. It is a small species, growing to about 20cm wide. It is a pale milky color and carnivorous. When confronted by danger, it will withdraw and burrow into the sand. The species is inhabited by the yellow clownfish (*Amphiprion clarkia*). This remedy is known partially from a case published in 2009 in InterHomeopathy by Marty Begin[125]. As such, this remedy remains tentative until it is further proved or triturated. This case is further complicated by the fact that Begin switched to Stictodactyla gigantea midway through his treatment of the patient. These two remedies do seem to be quite similar, leaving the question of how to distinguish them uncertain.

Key Issues: The main issue in this remedy, based on the limited info available, seems to be conflict. The being, as in Anthopleura, feels very dependent upon its social context for maintenance of the integrity of its self. However, its policy seems to be to simply give into the expectations and needs of others, rather than the more assertive and aggressive approach Stictodactyla takes. It withdraws from conflict, rather than engaging with it. Again this picture is very sketchy and should be taken as tentative until this remedy is further explored.

Mind: Timidity, fear of conflict. Unable to assert self.

Main Physical Issues: <bacon?

Main Physical Issues: Strong effect on the mucous membranes. Drainage of the external ear (followed by emerald or Hep Sulph). Headaches derived from toxicity. Yellowish green profuse mucous. Sensitivity of throat to the atmosphere. Catarrh. Nostril Ulceration. Laryngitis with green mucous. Affinity with Gallbladder Channel.

Clinical: Drainage of mucous membranes. Drainage of the respiratory tract

Miasm: ?

Color Preference: ?

Remedy Comparissons: Other Anthozoa, Other Cnidarians. Other Fossil Remedies. Emerald. Hepar Sulph. Sulphur. Pyrogenium. Agaricus.

Cases: Two cases are reported in Mercy's 2009 book. One was a female student with muscular pain and digestive issues overwhelmed by her classes. The second was a woman with a number of relationship issues.

Heteractis malu (hetera-ma)[124]:

This anemone, also known as the delicate sea anemone, is native to the Pacific, being found in Hawaii, Japan and Australia. It is a member of the Stictodactylidae along with *Styctodactyla gigantea*. It is a small species, growing to about 20cm wide. It is a pale milky color and carnivorous. When confronted by danger, it will withdraw and burrow into the sand. The species is inhabited by the yellow clownfish (*Amphiprion clarkia*). This remedy is known partially from a case published in 2009 in InterHomeopathy by Marty Begin[125]. As such, this remedy remains tentative until it is further proved or triturated. This case is further complicated by the fact that Begin switched to Stictodactyla gigantea midway through his treatment of the patient. These two remedies do seem to be quite similar, leaving the question of how to distinguish them uncertain.

Key Issues: The main issue in this remedy, based on the limited info available, seems to be conflict. The being, as in Anthopleura, feels very dependent upon its social context for maintenance of the integrity of its self. However, its policy seems to be to simply give into the expectations and needs of others, rather than the more assertive and aggressive approach Stictodactyla takes. It withdraws from conflict, rather than engaging with it. Again this picture is very sketchy and should be taken as tentative until this remedy is further explored.

Mind: Timidity, fear of conflict. Unable to assert self.

Main Physical Issues: <bacon?

Clinical:?

Miasm:?

Color preference: ?

Remedy Comparison: Pulsatilla, Silica, Anthopleura xanthogrammica

Stichodactyla gigantea (stich-g)[126]:

This anemone has a very large number of synonyms. It is also known as *Discosoma giganteum*, *Priapus giganteus*, Patchy giant anemone, Carpet anemone, Gigantic sea anemone, Giant carpet anemone, Indo Pacific sea anemone, *Anemone tapis*, and was originally proven as *Stoichatis Kenti*[127]. The name Stichodactyla gigantea seems to be the most biologically accepted at the moment, so it is the name chosen for this text. It is a large anemone (50-80 cm wide) which flourishes on sand ocean bottoms and seagrass meadows. It is symbiotic with algae species, depending on them for much of its food supply and dying without their presence. It is slightly mobile, and can move under threat. It is inhabited by several species of clownfish, anemone fish, and occasionally domino damselfish[128]. It was proven in 2004 at a spiritual retreat in Auroville, India by Nandita Shah and Melissa Burch after one of their patients identified at a source level with the sea anemone[129]. Jo Evans summarizes this remedy, but uses the latin name of a separate but closely related species *Stictodactyla haddoni*[130]. The proving information within Evan's book however derives from the same source, and so is likely accurate.

Key Issues: This being feels a lack of internal structure, which requires <u>support from outside itself to be maintained, particularly from the being's family, friends and partner</u>. Boundaries are lacking, and so <u>everyone in the being's support network must share the being's views and ideals</u>, particularly in the moral sphere. This is often phrased as <u>Integrity.</u> If members of the support network disagree with the being on something, or wish to go against the being's ideas of what should be, or their integrity, their sense of self is disrupted. They can react to this by either <u>withdrawal, or with irritability and rudeness.</u> This is verbalized quite well in the following paragraph from the proving:

> I feel a kind of determination to stop having relationships with people who have different opinions. Even if these people are important, I can step back if I have a different opinion. I fight my own truth without hate or compromise, even if I have to break the friendship. I feel more concentrated inside. I prefer people who pull me up then pull me down[131].

Mind: Antagonism with those around them. Feels choices are constrained by family/partner, lack of confidence, mistakes in writing and speaking. Stubbornness, intolerance of the opinion of others. Need for integrity. Rudeness. Timelessness. Difficulty with time. Withdrawal from others with different opinions.

Main Physical Issues: Accidents. Injuries. Sensations of upper respiratory tract infections, such as stuffy nose, coryza, sore throats, burning eyes. Sharp cutting, lancinating Pains. Eruptions on face. Extremity pain. Fatigue. Desires grapes, salt.

Clinical: Colds and other upper respiratory tract infections.

Miasm: ?

Color preference: ?

Remedy Comparison: Other cnidarians. Kali Bic, Calc-sil

Medusuzoa:
Chironex fleckerii (chir-f):[132]

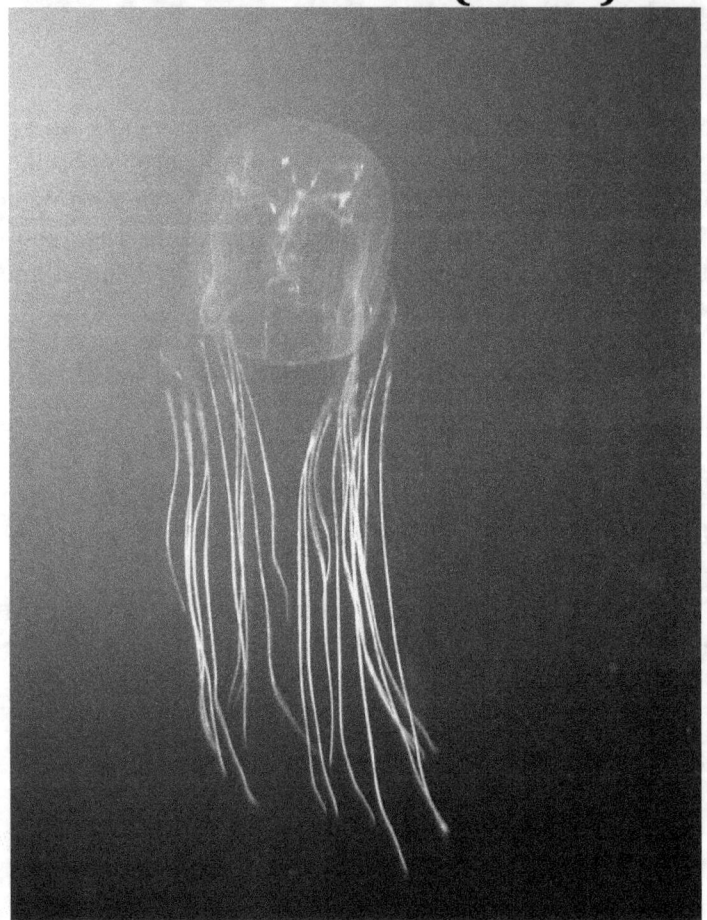

Also known as the Box jellyfish or sea wasp, *Chironex fleckerii* is a member of the Cubozoa, native to the tropical Pacific and Indian. This species in extremely poisonous, resulting in a large number of documented deaths. Stings produce excruciating pain and the venom has an LD50 of 0.4 mg/kg[133] (which is very low). Death can result within 3 minutes with extensive exposure. An antivenom exists, but the rapidity of the venom's effects prevents its effective use. Chironex possesses eyes almost as advanced as Cephalopods, giving it very good vision. It was proven by Alastair Grey at the Nature Cure College in Sydney in 2001 with a Hahnemannian methodology[134].

Key Issues: This remedy appears to have the core issue of order. The being feels a great need for order in life in order to maintain a sense of self and actively expends effort to create order to stabilize itself. Feelings of disorder or ambiguity will disrupt its sense of self and are thus opposed. Aggression will often be resorted to in order to restore a sense of order to the being's life. Breakdown occurs when the being is exhausted from maintaining order, or when the disorder is too much to cope with.

Mind: Desires order. Compulsion to cook and clean. Absent mindedness. Difficult to concentrate. Indecisive. Intuition increased. Sudden anger and aggression. Unrestrained sexual desire.

Main Physical Issues: Influenza-like illnesses. Throat pain, chemical burns. Better at seaside. Vertigo on standing still as if pulled forward. Coryza. Soreness of throat and tongue. Nausea/Vomiting. Affinity for calves. Affinity for forehead. Facial warmth and heat. Stinging and burning pain. Stomach as if punched. Strong hunger. Pressure on chest.

Clinical: Influenza? Upper respiratory tract infections?

Miasm: ?

Color preference: ?

Remedy Comparisons: Other cnidarians.

Medusa (medus, Aurelia aurita)[135]:

Aurelia is a small member of the Scyphoza. Interestingly, most members of this genus are extremely similar, and can only be differentiated with genetic testing, so our sample may in fact be of a closely related species. It is native to most of the world's warmer oceans. It is carnivorous, with a widely varied diet and thrives in conditions of low oxygen[136]. Aurelia is one of the oldest cnidarian remedies. It was proven by Houard in 1884 and by Gurge-Wickert in 1998[137]. Jo Evans summarizes these provings in her 2009 book, which is the primary source for this monograph[138].

Key Issues: The main issue of Aurelia is change. The issue of <u>change disrupts the integrity of their selves</u> and so they <u>expend a great deal of effort to create a very steady and consistent environment</u> for themselves. When the pace of change exceeds what they can cope with, they will experience dissolution of self.

Mind: Worse change. Teenagers coping with changes. Anger. Irritability. Hatred of others who enjoy life. Critical of others. Depression. Cut off others. Averse to company. Better alone, better eating. Worse company. Worse being asked questions.

Main Physical Issues[139]: Lack of milk production or alternately, excessive milk production. Urticaria, Skin inflammation, Vesicles. Leucorrhea.

Clinical: Skin rashes and vesicles. Herpes. Leucorrhea, Lactation issues.

Miasm: ?

Color preference: 15-16E

Remedy Comparisons: Other cnidarians, Urtica Urens, Rhus tox.

Cases: An excellent case of Medusa appeared in Links by Chauhan[140]. I would encourage readers to read it, as it gives an excellent overview of medusa from both a source, sensation and table of animals perspective (though Chauhan has not yet, to my knowledge, been aware of the table. It seems as if the themes emerged independently).

Physalia physalis (physalia-p)[141]:

Known in English as the Portuguese Man O'War, Physalia is a member of the Hydrozoa. It passively floats according to the tides and wind currents capturing prey as they become available. It is a colonial organism composed of individuals within several life stages of the same species, each of which with specific functions. Pneumaophores create the gas pocket which floats the colony, dactylozooids create the tentacles for defense and predation, gastrozooids for feeding and gonozooids for reproduction[142]. This remedy is also known as Physalia pelagica in Homeopathy, and is called by that name in Jo Evan's book[143]. There is not, to my knowledge, a proving or trituation of it. The symptoms derive from the poisoning symptoms described by Bennet in 1831[144]. I have a case of Physalia, written up below, though not as long a follow-up as I would like. The mental picture largely derives from this case.

Key Issues: The main issue seems to be <u>understanding</u>. An intellectual <u>understanding is required to make sense of the world and allow for a feeling of safety</u>. When this understanding is lacking the being feels immersed in great pain and loses the integrity of their sense of self. This is described as <u>stupidity or confusion</u>. Understanding preserves a sense of self, which is particularly disrupted by confusion and feeling stupid.

Mind: Mind heavy. Vagueness. Alienation. Dullness. Cannot concentrate. Impulsivity. Ailments from anger, rage, suppressed anger. Fighting with family and friends. Cannot think. Better occupied.

Main Physical Issues: Cold sensation inside head. Better in cold air. Better music. Blood sugar issues.

Clinical: Never well since Physalia stings[145]

Miasm: ?

Color preference: 15-16D, 15-16E

Remedy Comparison: Other cnidarians. Euplectella aspergillum. Badiaga.

Case:

The patient presented in January 2015 as a referral from another patient. He was a young man in his 20's with a main complaint of poor mental health. He had been recently hospitalized during one of his episodes, in which he felt a pain in his heart and was unable to speak. His workups at the hospital revealed nothing, and he was sufficiently lucid and capable to avoid being committed to a psychiatric in-patient facility. He had been prescribed Quentiapine (Seroquel), but had not taken it. He was on no other medications.

He reported being extremely emotionally labile since late December. He reported feeling very afraid, anxious and sobbing uncontrollably. He also reported very brief highs, lasting several minutes, in which it was impossible for him to feel badly about anything. Life seemed completely wonderful, These highs were always followed by days or weeks of lows.

He had recently begun experiencing blackouts. He reported a feeling of being foggy, of being unable to understand or perceive anything, followed by a period where he was unable to remember anything. These blackouts could last for days or even weeks. When they were over he regained memory and consciousness of himself, often not knowing where he was or what he was currently doing. This blackout appeared to not affect his behavior at all, as his family members reported no differences in his behavior when compared to times that he remembered. His prodromes were only a few seconds long, leaving him unable to warn friends and family members.

His previous history includes a diagnosis of ADHD, which was medicated previously. He reported the medication giving him a feeling of being a zombie, and discontinued it very soon after beginning. He was also prescribed psychiatric medication in the past that gave his a sensitivity to meat products upon withdrawal. He was a pescetarian, with a very good diet. He also reported very painful headaches occurring if he became excessively hungry.

On physical exam he displayed blood pressure of 100/64, somewhat low. His blood glucose was 5.0. His Chapman reflexes were positive on the Gallbladder, extremely positive on the pancreas and heart, positive on the stomach, positive on the peristalsis point, positive on the adrenal glands and mildly positive on the thyroid. He spoke extremely quickly, and jumped from topic to topic in such a way that grasping his narrative was somewhat challenging.

When asked to further describe the feeling he experienced right before his blackouts, he described a feeling of being in a fog. He felt stuck in this fog, as if something has a hold of him and is slowing him down. It was as if his brain was worried about something, and so went into this kind of slow survival mode to cope. He felt as if he was in a pile of mud, unable to move, think or do anything. When asked what he was afraid of, he remarked that he was afraid of not being able to understand. Any phenomenon, if understood, could be accepted, even if he didn't like it. Without understanding he just felt stupid.

When asked to describe what being stupid was like, he described his normal state. Normally, he saw an object and understood it. He understood how a car worked, how walls were wired, etc. Being stupid was a state in which his brain stops, does not let in information, and dumbs itself down. It was a defensive mechanism; he felt it helped to prevent things from being able to hurt him.

When asked what was hurting him, he jumped to his relationship with his mother. He then jumped to other relationships he had, where people had hurt him for no reason, just for the sake of hurting. I asked him to describe the pain and he labeled it as a global pain, affecting all of his body at once, no part hurting more or less than another. It was like being dipped in acid, and during this pain his brain wasn't working.

He then went very thoroughly into his state. He described the pain as everything being over stimulated. Everything in his perception overlapped. It was as if someone dunked him in acid. There were alternating sensations of hot and cold, and no part of his body worked. During these periods he needed a huge amount of energy to do anything. Simple tasks like thinking or moving seemed to take forever, and were excessively difficult. He also reported that during these pain periods his perceptions were all in black and white.

He described himself as floating passively in this pain. Nothing was holding him up. He felt buffered by a current, but not held up by it, a sensation that puzzled him.

The opposite of this pain was sleep. In sleep he was able to build a world which nothing would interfere with. He could make his world logical and understandable, this was his favorite aspect of dreaming. He also perceived in full color here. He was an accomplished lucid dreamer, creating many wonderful novels out of his nightly adventures.

I asked the patient whether it was confusion or ambiguity which truly triggered his pain. He responded to being fine with ambiguity, but that confusion was the main issue.

To summarize the patient's main issue was his understanding of the world around him. He suffered from episodes (triggered by blood sugar or brain neurochemistry, or quite possibly both) in which he was unable to process and understand the world. This led to a global sensation of pain, like being immersed in acid. Accompanying the pain was a feeling of being free floating and being buffered by currents, though not held up by them. The opposite of this pain was to be found in lucid dreaming, when he could create new and logical worlds. His color preference was 15-E.

This patients issues with global sensation place him very firmly in the cnidarian group, but I was uncertain which remedy he belonged to. Having recently finished the Cnidarian chapter, I was certain the patient was not any of the known Cnidarian remedies. The lack of a withdrawal reaction did suggest a Medusuzoa. His source information of passive floating however suggested very clearly a remedy. I prescribed Physalia physalis 1M 1 drop taken per day succussed.

I also prescribed Unda 34, 10 drops twice a day to deal with his blood sugar issues along with regular mealtimes and avoidance of snacks. If he began having headaches, he could have a drink of juice. I also prescribed 5 ml of Swedish bitters, to be taken and held in the mouth for 20s before meals, to stimulate digestive secretions.

The patient reported at his next follow-up a strong amelioration of all symptoms. He reported particularly lucid dreams on nights he took the remedy, sometimes interfering with his ability to feel fully rested. His blackouts had reduced considerably and he reported a general improvement of his mood and condition. The sensation of passive floating had diminished considerably as well.

Over time this patients work demands took him away from my practice. He eventually decided felt homeopathy was not the treatment method for him, after reading a number of "skeptical" articles about it on the internet. His dramatic response to this remedy however did motivate me to include it within this book.

Summary of the Key Issues Of Cnidarians:

Remedy	Main Issue Within The Table
Anthozoa	
Anthropleura xanthogrammatica	The being feels profound connection with others which stabilizes their sense of self. Anything that disrupts their connection to others, especially taboo and unacceptable behaviors, or social conflict, disrupts their connection and their sense of self.
Corallium nigrum	Social conflict and having others make social demands which tax the beings inner reserves, disrupts the being's sense of self. The response is to withdraw from conflict and isolate oneself.
Corallium rubrum	The being feels their sense of self dissolving because of the intensity of sensory stimulation the receive from the external world. They will respond by hiding and withdrawing.
Diploria clivosa	The being requires symbiosis with lifeforms around it in order to maintain the integrity of its self. This symbiosis is disrupted by chaos, brought on by willfulness, going against the higher divine will.
Fossil Dimorphastrea	The being feels extremely intoxicated by its environment, both environmentally and socially. It withdraws into seclusion and isolation. It perceives this isolation as lonely however, wishing to have lively exchange with the outside world.
Fossil Fungia coral	The being is immersed in its ancestral inheritance so completely that it is unable to completely live its own life. It floats above all involvement with its life, and is unable to fulfill the roles of its current life, or even truly feel it.
Heteractis malu	Conflict disrupts the beings sense of self, so it withdraws and gives in to the demands of others to avoid it.
Stichodactyla gigantea	The being requires support from it's social circle to maintain their sense of selves. It requires others to adhere to their moral codes and have their sense of self is disrupted when others aren't acting according to the being's ideals. The being usually perceives this as acting against their own personal integrity. They will either become quite irritable or withdraw in response to this.
Medusozoa	
Chironex fleckerii	Disorder disrupts the integrity of the beings sense of self, so the being actively works to create order in its life. Aggression is often resorted to in order to restore order. Dissolution occurs when the being is exhausted an unable to order its world, or it is surrounded by to much disorder.
Medusa	Change disrupts the beings sense of self-integrity, and so they actively create a stable environment for itself. If change exceeds the beings ability to cope, it goes into crisis and feels itself dissolve.
Physalia physalis	The being feels that it needs to maintain an understanding of the world around it. Confusion and feeling stupid destroy its sense of self.

Ctenophores:

Figure 10: *Mertensia ovum*[86]

The ctenophores, or Comb Jellies, are a small phyla of about 150 species. They are defined by having rows of fused cilia, and by adhesive prey capturing cells called colloblasts[146] and are also almost universally bioluminescent[147]. They are found worldwide in marine environments. A photo of an example species, *Mertensia ovum*, is shown right[148]:

The cilia are the most distinctive feature of the ctenophores. There are usually eight rows of them, arranged running from the mouth bearing end of the organism (oral) to the non moth bearing end (aboral). Cilia are located in small groups in their row, called a ctene[149]. These cilia can be remarkably iridescent. Unlike cnidarian medusae, ctenophores propel themselves by coordinated beating of the cilia, rather than by muscular contraction. They beat in an organized wave, starting at the oral end, ending at the aboral ending. Thus ctenophores tend to swim mouth first.

In addition to cilia, most ctenophores (outside of the smaller nuda class) also possess tentacles, large structures similar to those of other phyla, which primarily function to capture prey with colloblast cells. Colloblasts are small coiled filaments that are coated with small eosinophilic granules[150] located on the tentacles. When triggered they spring and adhere onto prey while the tentacles are drawn back into the mouth for feeding. Unlike cnidocrytes, these cells are not venomous, being merely adhesive. A photo of an unknown species of ctenophore with prominent tentacles is displayed below[151].

Bioluminescence is another widely shared characteristic among the ctenophores. Light is produced by calcium activated photoproteins, similar to luciferins in other species, in specialized cells called photocytes[152]. The purposes of luminescence seem to be highly varied. Communication, mating, camouflage and cellular signaling have all been suggested as roles for this phenomenon, but final conclusions have yet to be made[153].

In terms of general biology, ctenophores are quite similar to the cnidarians. Their bodies are also composed of two layers of epithelial cells separated by a thick jelly like mesoglia.

Epithelial cells are linked together in a basement membrane, and share intracellular connections. The outer epithelium secretes a mucous coating that protects the organism, and also contains stem cells.

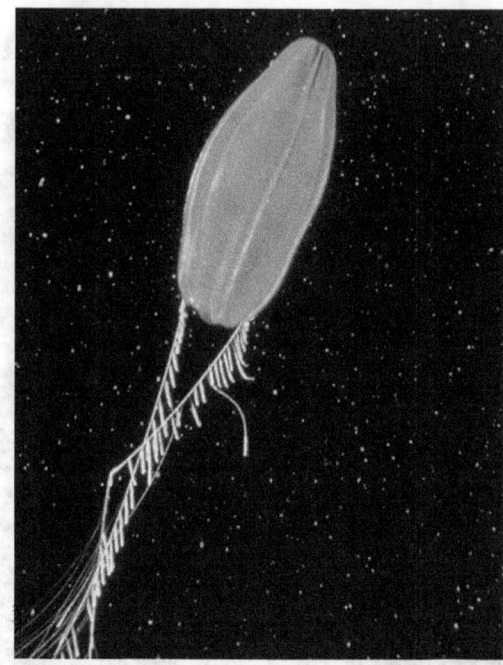

Figure 11: Unknown Ctenophore[89]

The digestive system consists of a tube through the organism, sealed on either end by a ring of small muscles and a small area in the center of the organism where digestive enzymes are released, functioning effectively as a stomach and intestines. The anal pore does not function primarily to excrete, with most unwanted ingested matter being regurgitated from the mouth.

No central nervous system is present, although a neural net exists, and is particularly concentrated around critical structures such as the cilia. One sensory organ[154] is usually present containing a solid particle surrounded by nerve cells, which appear to sense the pressure the solid object applies on them due to gravitational pull. It functions as a position sensory, allowing the organism to sense its orientation, and take corrective actions.

The basic body plan of a cydippid ctenophore is displayed diagrammatically below[155]

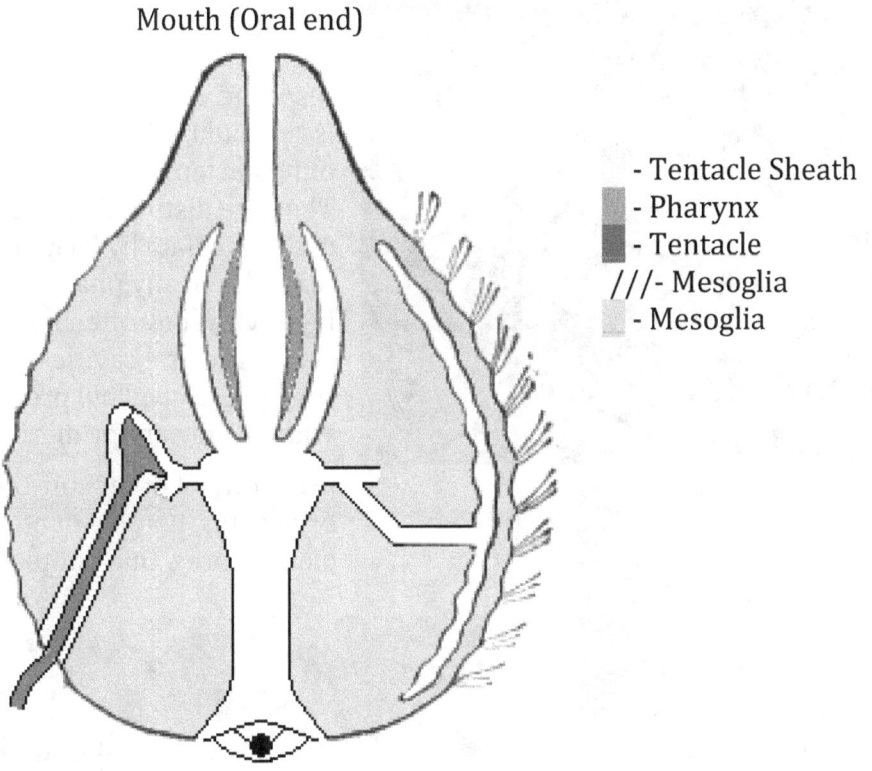

Most ctenophores are hermaphrodites, and self-fertilization does often occur. Most species externally fertilize by releasing gonads into surrounding water, but some species in the platyctenids have internal brood chambers which house fertilized eggs until hatching.

Figure 12: *Coeloplana astericola*[94]

For such a small phylum, ctenophores display considerable diversity. While a full look at Ctenophore diversity is beyond the scope of this book, we will examine the three known classes. The first and largest is the tenticulata which is defined by the presence of tentacles which can be retracted into the body and which are used to capture food. One particularly intriguing order within this class is the platyctenida, an order which resembles an aquatic worm, living largely benthic lives on the sea floor, and which use brood pouches to reproduce. They're also notable for coming into a symbiotic relationship with many echinoderms. A photo of one species, *Coeloplana astericola*, is shown above on top of a sea star species[156]:

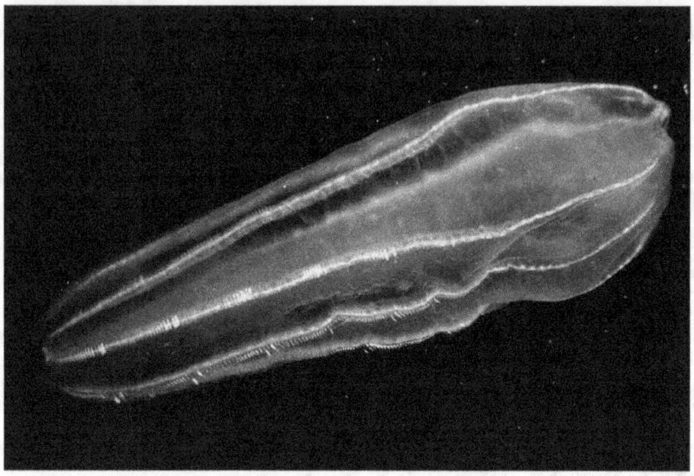

A second class of ctenophore is the nuda, which contains only one family, two genera. They are distinguished from the tenticulata by lacking tentacles in any phase of their life cycle. They often have very large oral cavities, which they use to swallow prey whole. They are primarily predators of other soft-bodied organisms, notably of other ctenophores. One photo of an unknown species is shown above[157]:

Figure 13: A Nudan Ctenophore[95]

The final class of ctenophores is extinct, and known solely from some fossils in Yunnan. The cleroctenophora are dated to the early Cambrian, and are unique in the possession of a skeleton that supported the body[158]. One drawing is shown right of *Maotianoascus octarius*[159]:

Ecology of Ctenophores

Ctenophora have a remarkable role in the ecology of the world's oceans for such a small phyla. Found globally in most depths and in all temperatures of ocean, most ctenophora are active predators, feeding on plankton captured from the surrounding waters and sometimes larger organisms[160].

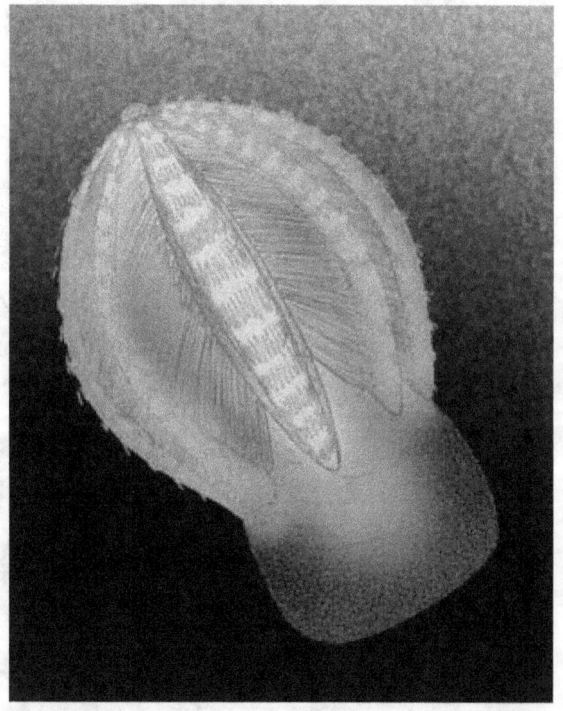

Figure 14: Maotianoascus octarius[96]

They can consume through large amounts of plankton, sometimes upto 10 times their bodyweight. Their prey can range from small zooplankton to larger animals such as cnidarians and other ctenophores. A variety of feeding strategies are used, from simply passive waiting for phytoplankton, to active hunting.

Ecologically, ctenophores are believed to generally act as controllers of copepods and thus maintainers of the algal populations that copepods feed on.

Evolution of Ctenophores:

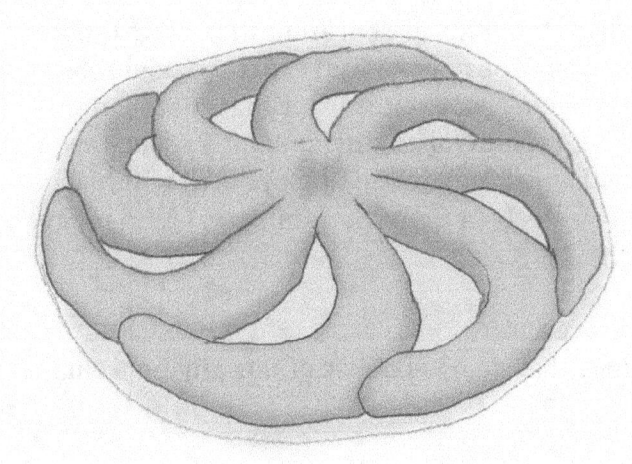

Figure 15: *Eoandromeda*, a possible Ctenophore ancestor[101]

The evolution of ctenophores is very difficult to track due to the lack of hard body parts amenable to fossilization. However, some early ctenophores appear to have possessed bony parts, leaving some fossils dating from the early Cambrian[161]. Other fossils have been found from the stephen formation in British Columbia dating to the mid Cambrian[162] suggesting that large branches of this phyla may have become extinct.

Many fossils of ediacaran organisms resemble Ctenophores. One example shown in figure 13 is that of the creature *Eoandromeda*[163], a creature which may have been an ancestor of ctenophora[164].

Many ctenophore-like fossils have been found that are cessile animals that appear to use cilia to filter feed[165]. Shu et al suggest that such creatures are relatively close to the ancestors of ctenophores, and that this phyla originated as largely sessile animals and that a free floating lifestyle was a later adaptation[166]. An example of these creatures, *Stromatoverus psygmoglena*, is shown in figure 14[167]

Given the limitations of the fossil record it is perhaps better to rely on genetic data to determine the relationship of ctenophores to other phyla. Like sponges, ctenophores lack Hox genes, a set of

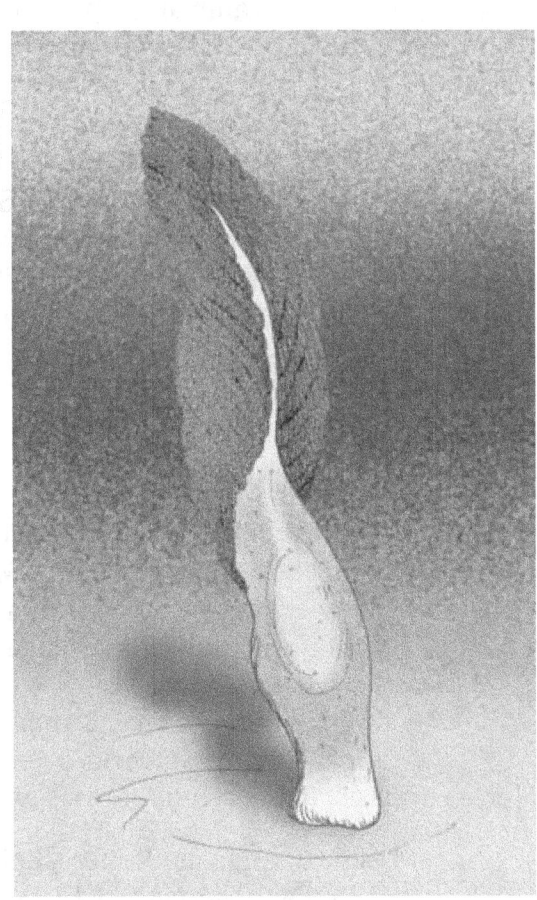

Figure 16: Stromatoverus psygmoglena[105]

developmental genes [168]. They also lack such fundamental cellular mechanisms as zinc fingers[169], microRNA[170] fundamental cellular signaling pathways such as the WNT path[171] and massive differences in mitochondrial DNA from other presumably nearly related phyla[172]. Neural chemistry and signaling is also radically different from the rest of the animal kingdom, suggesting that ctenophore may have developed neurological systems completely independently from the rest of animal life[173].

Two competing hypotheses have come to preeminence over the interpretation of these findings, the first of which is that Ctenophores represent the first extant branching of the tree of animalian life, and the sponges have lost considerable amounts of their once more complex cellular machinery[174] or that cnetophora evolved such things as muscles and neural networks independently of the rest of the metazoa[175]. Because of the Homeopathic position of the Ctenophores within layer 1, and the absolute uniqueness of many aspects of their biochemistry, I tend to favor the latter explanation.

Unique Characteristics of Ctenophores:
Ctenophores have such unique biology that many features of this phylum remain unknown outside of it.

1. Cilia: Ctenophores are the only known animal phylum to move via Cilia on a macro (rather than micro) scale. This means of propulsion is unique within the animal kingdom outside of microorganisms.
2. Coloblast cells: Also unique to the ctenophores are colloblasts, adhesive sells which adhere to prey animals, assisting in capture.
3. Fundamentally different cell signaling: Both intra and extracellular communication systems among ctenophores display vast differences from those of the remainder of the animal kingdom, with many completely novel mechanisms and chemicals being used instead of the more familiar substances and processes in higher animals.
4. Bioluminescence: Ctenophores are almost universally bioluminescent, through the exact reason for this has yet to be discovered.

Ctenophores in Homeopathy:

Unlike the cnidarians, known in homeopathy since Hahnemann's time (and well before in folk medicine), ctenophores have no history of use within either folk medicine or in homeopathy. My trituration of *Mnemiopsis macrydi* is to my knowledge the only homeopathic exploration of this phylum that exists. As such this information must be taken as tentative.

Ctenophores are layer 1 remedies, with a major focus on incarnating and entering into the manifest world. From my experience with *Mnemiopsis*, I posit that the main issue in the ctenophores is experience and the types of experiences. They are very simple beings, focused on sensory experience and not thinking or reflecting on those experiences greatly. They prefer to drift through life passively, experiencing whatever may come, without expending any effort to direct those experiences.

However, many of those experiences are very unpleasant and cause the being great distress. Though we have only one trituration, I suggest that it is this reaction to unpleasant experiences that will differ between ctenophore species. *Mnemiopsis* reacts with fear, feelings of insecurity and a feeling of danger. If my hypothesis is correct, other species will display different reactions.

The unpleasant experiences that the being experiences can be avoided by taking an active role in the direction of its life. In *Mnemiopsis* the being perceives the directing of its life as profoundly energy draining, and develops great fatigue if it should attempt to do so. Other ctenophores may display differing reactions to the need to actively direct their lives.

True healing for people needing ctenophore remedies is in finding experiences they genuinely desire. This animates them, giving them the energy and motivation to direct the course of their lives actively.

Fortunately for those who practice sensation method, a number of very clear sensations emerged from the trituration which can now be used for this style of casetaking. Prominent sensations are:
- Dreamy
- Spacy
- Feeling of drifting
- Sensations of flowing water, within ones body and around it
- Fluid
- Formless
- Passivity
- Feeling of active movement towards goals associated with a sensation of shivering
- Floating
- Jittery
- Feeling like Jelly
- Squishy
- Feels like a plastic bag full of water
- Feels like a water balloon

A colleague, Dr Ghanshyam Kalathia has written the following text upon reading my trituration of *Mnemiopsis macrydi*. His contribution is gratefully acknowledged.

Sensation of Ctenophora- Dr Ghanshyam Kalathia

Support system

Ctenophores are more advanced than porifera and similar to cnidarians. They have a tissue level of organization, two layers together that form the structure of the animal. Like cnidarians they are stuck at an infant stage, but they also share some expressions similar to the porifera.

They are newly formed and so seek active support. They need support to complete activities of daily living such as feeding and taking care of the self. They have just become aware of own self and of their own body, so they are very much afraid to lose their simple form. They feel defenseless and weak without support. Even the simplest effort is too much for them. They get exhausted very quickly.

Their newly formed structure is not stable, so they need support that can stabilize them. Without support they feel extremely vulnerable and fearful. The world becomes musty, unclear and they become dreamy or spacey. In this state, they are unaware of the self as well as the world. Sometimes patients might say that suddenly they lost their sight and everything becomes foggy.

Simple form and structure

Their simple tissue level of internal organization makes ctenophores among the most basic of animals. They have awareness of their own form but it is so simple that they just perceive their experiences rather than reacting or thinking much about them. Their structure is so simple that it can break during the smallest adversities and create turmoil in them. They try very hard to keep this structure intact. The simplest activity is too much for them, they get drained out very easily, like climbing mountain without legs.

They love to be whatever they are rather than to make big changes in themselves. They love to be simple and naïve. They do not like any kind of complexity. They try to flow with the situation rather than modify it or make an effort against it. They do not have capacity to drive against any force. So, they simply accept and flow with outside situations and direct themselves wherever they drive them. But this flowing attitude makes them completely lost and spacey. They forget their own reality and so everything becomes unreal or misty.

Unstable

When they go in to uncompensated state, they become totally unstable. Because of instability they have an extreme fear of losing their balance. In this context they are very similar to the cnidarians. They try to stay in balance, so their main focus is to keep their form and structure. Funnily enough cnidarians and ctenophores both have specialized cells called 'statocyst' which are balancer cells.

Instability also creates changeability of emotions, attitude and mood. Their entire being has no base and nothing is steady; everything is changing and alternating. They drift according to the situation and it creates irresolution. They do not think much but react by intuition or instincts.

Spacey

The ctenophore pattern has a polarity of drifting away or flowing along the situation and just experiencing whatever comes up. The shadow side of this pattern is that they get lost and are not able to re-collect thier own real personality. This is very troublesome for them because it takes away thier newly formed awareness of the self.

They also feel they are not capable of thinking any more. They do not have thoughts. They feel a freezing in the brain. On the other hand, when they feel their brain is fine they get lost in the mental realm and start to create vivid images. They like their vivid perceptions and their own fantasy world.

Vulnerable

When the patient is well compensated then they can either drift out or become spacey. Uncompensated cases are much more vulnerable. Their vulnerability is regarding the losing of their structure. This can be extremely fearful and makes them completely unstable. Patients might have panic attacks or might have phobias. They need to hold on whatever they gain, so they make a huge effort in regard to it and finally become drained out and become exhausted. Mostly they have unqualified fears. They have fears but are not able to explain of what and why they are afraid.

Desire for experiencing

Like cnidarians they have a newly formed structure; they are like a newly born baby. They are amazed by everything and try to experience everything. "Feeling" is the main focus for them. They try to sense everything. For them every sensation, feeling and experience is a huge pleasure. Ctenophores are more passive then cnidarians, so here we miss the reactive counter part of the experiencing process Here they like to drift in the experience rather than go into any kind of process.

Reactions – Passive – No reaction only Perceptions

Cnidarians receive stimuli and then react, while ctenophores perceive everything display less reaction. That's why they look very similar to porifera, but in sponges the passivity is their counter polarity. Their main focus is to be there, to prevent their existence. So, either they try to put effort towards keeping their existence or are passive. The ctenophores either try to drift out and experience passively or become spacey and exhausted. The main focus of the Ctenophores is to keep their form intact, but the sponges have no form and so they try to keep within existence.

Luminescence
Luminescence is common characteristics of all ctenophores. It expresses in the cases as desire for colors, image or dream of reflected colors, reflected light, auto-reflection etc.
Gluey, Sticky, Trapping

As a defense ctenophores have gluey, sticky thread like tentacles which entrap preys. This express in cases as gluey, sticky, entrapping, catching, and adhering sensations.

SUMMARY OF SENSATIONS

Specific Features
Tissue level of organization, activities are coordinated by a decentralized nerve net and simple receptors, radial symmetry. Comb – group of cilia for swimming. Colloblasts – sticky tentacles to catch the prey. Fragile body. Photocytes that produce bioluminescence. Sexual and Asexual types of reproduction

Support Systems and Dependency
Seek support actively
Need support to keep balancing
Need support for daily simplest activity
Need support to keep new form intact
Dependent like an Infant

Sensations
Simple form
Thin skinned (plastic bag, water filled balloon)
Vulnerable, fragile, breaking easily
Unstable, unbalanced
Sensual, amazed by experiencing and perceiving passively
Panic, fear, fright

Physical sensations
Drying-out, breaking, belting, drain out, weakness, collapsing, exhaustion

Reactions
Extremely fearful
Passive and just drift according to situation
No reaction, only perception
Exhaust out, drain out

Defenses
Be passive
Glue out, stick out

    Entrapping, catching, adhering
    Luminescence
Physical affinities
    Phobias, panic attacks, bipolar disorders, behavioral disorders, skin complainys, developmental problems, chronic fatigue syndrome

Suggested Provings of Ctenophores:

Ctenophores are so far almost completely unknown within homeopathy, with only my own trituration proving existing. As such a proving of any ctenophore would be a very useful addition to the homeopathic materia medica. Since my trituration was of a tenticulata, a proving of a member of the nuda would be the best use of effort.

One interesting species of ctenophores for proving would be *Haeckelia rubra*, a species which capture the cnidocytes of ctenophores. Due to their unusual benthic and symbiotic lifestyle and wormlike bodies, members of the *Platyctenida* would also be excellent candidates for proving.

Mnemiopsis macrydi[176]

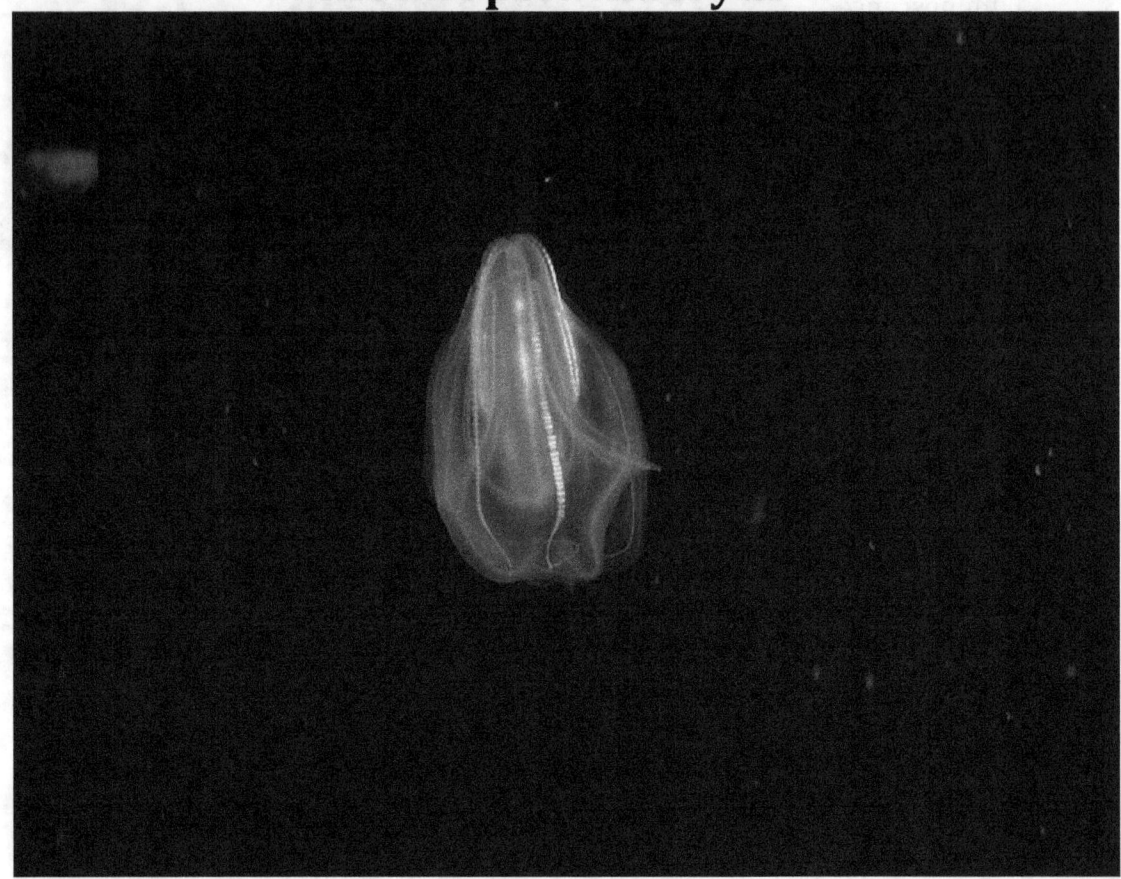

Known as the warty comb jelly or the sea walnut, Mnemopsis is a member of the lobata order of the tentaculata. It has 4 rows of Cilia combs. It feeds upon on small zooplankton. This individual was obtained from a marine specimen laboratory and triturated by myself alone. I drowned the specimen in vodka, then triturated a small portion including cilia and mesoglia.

Key Issues: The key issue in this remedy seems to be fear and danger. This being is immersed in sensation and perception. It drifts through life, experiencing these perceptions and gaining from them. Its natural inclination is to drift and simply passively experience, but it is greatly troubled by the disturbing things it experiences. It feels great danger and a lack of safety, and it experiences great fear. It can avoid this danger and fear by actively directing its course in life. However it finds that directing its course takes great energy, and it becomes fatigued from the effort. Healing for this being involves finding experiences it actively wants to have, and using the energy of that desire to fuel its efforts to direct its life towards those experiences.

Mind: Dreamy. Spacy. Vivid perception. Difficulty thinking and planning. As if intoxicated. Feeling as if life is simple. Feeling of exhaustion when contemplating taking action toward a given outcome. Passivity. Helplessness.

Physical Symptoms: Sensation of cool water running down my shoulders. Fatigue. Jelly squishy feeling throughout body.

Clinical: None known.

Miasm: None known.

Remedy Comparison: Other layer one remedies. Hydrogen. Porifera.

Cases: None.

Mnemiopsis macrydi Trituration C1-3

The initial levels of the trituration present the picture of a being immersed in sensation and perception. The being drifts through life, experiencing things on a purely sensory level, not thinking about its experiences or reflecting on them. It lives a simple life, going where the currents of life draw it. However the being experiences much in life that is unpleasant, and specifically very frightening. It feels great danger in the world and a profound lack of safety, but in order to avoid this danger the being must actively direct its course throughout life. This feels like a huge amount of effort to the being, exhausting it. The being is caught between frightening experiences of perception and a state of exhaustion from trying to direct its course away from those frightening things.

C1:
Preformed October 8th 2015:
- I feel dreamy, spacy and almost a bit unreal
    - But my experiences are also very vivid!
- There is a feeling of cool water rushing down my shoulders
    - The sensation in that area seems a great deal more vivid to me
- I feel very dreamy, it is hard to think clearly
- Everything is very immediate, very vivid. It is as if there is nothing in my perception but my immediate sensation
    - I feel very afraid right now
- I am more removed and dreamy…. I'm getting words mixed up
- I feel very sleepy
- I only perceive. There is not a lot of reaction to my perception
- I feel very fluid and formless, just impressions and perceptions
- I literally feel a freeze in my brain. I cannot think or process!

C2:
Preformed October 12th 2015:
- Again I feel the lack of thought and processing, as well as the vivid sensation
    - There is not much thinking, just perceptions
- There is a certain joy in living this way. It feels wholesome, organic, simple. I get a real sense of simplicity. An idea occurs to me " why would I want to make things complicated?"
    - It feels almost as if I am high. Perception is very captivating
- I'm forgetting to do things, daydreaming, getting lost easily
- There is a deep calmness to this remedy. Everything is very simple when you let it be so. Just observe it, don't think much about it!
- I have a sense of moving through flowing water. I'm drifting, going wherever the water takes me
    - I have no active direction, there is no struggle to get anywhere
- Complete surrender, as the mystics say. There is no sense of identity to direct my course. I merely drift and experience
- I'm so at peace. I have no problems
- I am now somehow uncomfortable. There is a feeling of being helpless and just suffering. I have no capacity to react
- What I am perceiving disturbs me. Where I am going is uncomfortable
    - Ahhhhh. I suffer!!!

C3:
Preformed October 14th 2015.
- I am troubled. I'm bothered by something but cannot quite put my finger on it
    - I just cant quite conceptualize it
- I am disturbed. I feel a vague sense of unease
- I'm scared, distressed and helpless. I can't do anything about it
- I feel like I am so passive. I can only just react to things I perceive. But I have no control
    - In order to change things I have to do more than just drift
- I need to do more than just drift and react and perceive. I need to somehow actively engage, to actively do things. I would need to take control of my experience
    - But I am afraid to do so!
- It is so much effort. Such energy is needed to actively direct oneself, to engage with the world! Its so much easier to drift and experience
- I'm stuck between the possibility of this unpleasant experience and between the work it would take to direct my experience actively
    - I can't possibly sustain that level of energy
- I'm troubled by both options I have
    - To be or not to be active in my choices

C4

In this level of the trituration the being feels a sense of safety. It now feels safe, and unafraid of its experiences. With this new feeling of safety, the being begins to experience a sense of curiosity or a desire for experiences. For the first time there are things that the being actively desires to experience, and this desire gives the being a burst of energy, alleviating the sense of exhaustion felt in C1-3 when the being contemplates actively intervening in its path in life. This energy gives the being the stamina to actively direct its life. A sensation of shivering was associated with this active posture.

- I feel good. I am very happy peaceful and joyous!
- I feel safe. Safe enough to move around on my own now. Somehow what I am experiencing is less frightening now
    - I feel curiosity now. I think I will be actively seeking perceptions now, but not out of fear of some perceptions, but out of a desire to experience certain things
- I want certain experiences. I will seek them out!
- I feel a sense of excitement about all of the things I am going to experience now. If they don't appear to me I will and seek them out!
- This excitement, this zest for life seems to give me some sort of energy to actively seek out things. I am no longer fatigued!
- I pleasantly await the experiences I will have!
- I feel like I am floating and enjoying my self. I can now <u>shiver</u> and direct myself where I want to go. I can direct my own experience
- I feel excited!
- I feel a jittery jelly squishy feeling throughout my body
    - But this feeling is not in my heart. That feels solid for some reason
- I feel like a plastic bag full of water, or a very loose water balloon
- Again a feeling of shivering and propelling oneself

C5

In this level of trituration we return to the fears of the nature of one's experience found in C1-3. The being feels as if it is embarking on a great journey, but does not know what it will experience. Its natural inclination is to drift through life without active engagement with what it will experience, but this provokes great anxiety about this lacking any control over the content of those resulting experiences. Active engagement with life and selection of experiences would solve this problem, but is very exhausting to the being.

Preformed October 15th 2015:
- I feel nervous again and troubled

- I am worried about something
- What will my experience hold?
- I feel like I am starting on a long journey but that I do not know what it will entail
- What will I experience? What will I do? How will my life turn out?
    - And what if my path turns out badly?
- I am afraid of the potential unpleasantness I might face
- I might not want to do this. I might just want to avoid the whole enterprise
- The issue is the type of experience I will have. Will this experience be what I am looking for, or what I need?
    - If I continue passively drifting, I don't think it will
- In order to have the type of experience I need I am going to have to move out of passive reception and actually take control of life
- I'll need to actively do things for myself rather than just drift and perceive without taking an active role
    - But it just seems so exhausting!
- I would just rather passively experience than actively direct my course of life. Its all about passivity
- It is a decision to take up with life or to let life happen to me and take no action on anything
- I feel a great fear and uncertainty
- I need to move actively but it is so difficult and frightening to do so!

---

[1] Lévi, C. 1999. Sponge science, from origin to outlook. Memoirs of the Queensland Museum Volume: 44 Part: Year: 1999 44:1-7.

[2] Lavrov, D. Porifera. Tree of Life Web project. Accessed October 12th 2014. http://tolweb.org/Porifera/2464

[3] Image taken from Wikimedia commons. http://en.wikipedia.org/wiki/File:Porifera_cell_types_01.png

[4] Lavrov, D. Porifera. Tree of Life Web project. Accessed October 12th 2014. http://tolweb.org/Porifera/2464

[5] Anonymous. Porifera: The Cells. Online Document Accessed Oct 12th 2014. http://www.ucmp.berkeley.edu/porifera/pororg.html

[6] IBID.

[7] IBID.

[8] IBID.

[9] IBID.
[10] IBID.
[11] Kelvinsong. Image taken from Wikimedia Commons.
http://commons.wikimedia.org/wiki/File:Sea_sponge.svg
[12] Anonymous. Porifera: Life History and Ecology. Accessed October 12th 2014.
http://www.ucmp.berkeley.edu/porifera/poriferalh.html
[13] IBID.
[14] IBID.
[15] Lavrov, D. Porifera. Tree of Life Web project. Accessed October 12th 2014.
http://tolweb.org/Porifera/2464
[16] Anonymous. Porifera Tree of Life. Online Document Accessed Oct 12 2014.
http://www.onezoom.org/porifera.htm A part of the Profera Online Database.
- Van Soest, R.W.M; Boury-Esnault, N.; Hooper, J.N.A.; Rützler, K.; de Voogd, N.J.; Alvarez de Glasby, B.; Hajdu, E.; Pisera, A.B.; Manconi, R.; Schoenberg, C.; Janussen, D.; Tabachnick, K.R., Klautau, M.; Picton, B.; Kelly, M.; Vacelet, J.; Dohrmann, M.; Díaz, M.-C.; Cárdenas, P. (2014) World Porifera database. Accessed at http://www.marinespecies.org/porifera on 2014-10-12
[17] Anonymous. Porifera: The Fossil Record. Online Document Accessed Oct 12 2014.
http://www.ucmp.berkeley.edu/porifera/poriferafr.html
[18] Pechenik, J. Biology of the Invertebrates. 2000. Mcgraw Hill. Toronto. Pp. 82-3.
[19] IBID.
[20] Hershman, D. Online image Accessed Feb 13th 2015.
http://commons.wikimedia.org/wiki/File:Aphrocallistes_vastus-_dphershman.jpg
[21] Pechenik, J. Biology of the Invertebrates. 2000. Mcgraw Hill. Toronto. Pp. 82.
[22] SIMoN/MBNMS. Online Image accessed Feb 13th 2015.
http://upload.wikimedia.org/wikipedia/commons/e/e4/Acarnus_erithacus.jpg
[23] Pechenik, J. Biology of the Invertebrates. 2000. Mcgraw Hill. Toronto. Pp. 80.
[24] Borchiellini, et al. "Sponge paraphyly and the origin of Metazoa". *Journal of Evolutionary Biology* **14** (1): 171–179.
[25] Esculapio. Online Image Accessed February 13th 2015.
http://en.wikipedia.org/wiki/File:Clathrina_clathrus_Scarpone_055.jpg
[26] Anonymous. Calcarea. Online Document Accessed Oct 12 2014.
http://www.ucmp.berkeley.edu/porifera/calcarea.html
[27] Anonymous. Desmospongia. Online Document Accessed Oct 12 2014.
http://www.ucmp.berkeley.edu/porifera/demospongia.html
[28] Anonymous. Hexactinellida. Accessed Oct 12 2014.
http://www.ucmp.berkeley.edu/porifera/hexactinellida.html
[29] Anonymous. Porifera: The Fossil Record. Online Document Accessed Oct 12 2014.
http://www.ucmp.berkeley.edu/porifera/poriferafr.html
[30] IBID.
[31] Anonymous. Phylum Porifera: Sponge Fossils. Online Document Accessed Oct 12th 2014. http://www.fossilmuseum.net/Tree_of_Life/Phylum-Porifera.htm
[32] Evans, J. Sea Remedies: Evolution of the Senses. 2009. Emryss. P153.

[33] Jaworski, H. The Sponges-Their Significance. British Homeopathic Journal. 49(2). April 1960. Pp. 121,279. Cited in Evans, J. Sea Remedies: Evolution of the Senses. 2009. Emryss. P153.
[34] Anonymous. Porifera: The Fossil Record. Online Document Accessed Oct 12 2014. http://www.ucmp.berkeley.edu/porifera/poriferafr.html
[35] Anonymous. Porifera: Life History and Ecology. Online Document Accessed Oct 12th 2014. http://www.ucmp.berkeley.edu/porifera/poriferalh.html
[36] IBID.
[37] Wagner, M. & Behnam, F. The Ecology of Marine Sponge Associated Organisms. Online Document Accessed Oct 12th 2014. http://www.microbial-ecology.net/sponge.asp
[38] Bergquist, P. R. Porifera (Sponges). Encyclopedia of Life Sciences. 2001. John Wiley & Sons, Ltd.
[39] Keesing, J. Usher, K. Fromont, J. First record of photosynthetic cyanobacterial symbionts from mesophotic temperate sponges. 2012. Marine and Freshwater Research 63(5) 403-408.
[40] Mehbub et al. Marine Sponge Derived Natural Products between 2001 and 2010: Trends and Opportunities for Discovery of Bioactives. Mar. Drugs 2014, *12*, 4539-4577.
[41] Anonymous. Porifera: Life History and Ecology. Online Document Accessed Oct 12th 2014. http://www.ucmp.berkeley.edu/porifera/poriferalh.html
[42] Van Soest RWM, Boury-Esnault N, Vacelet J, Dohrmann M, Erpenbeck D, et al. (2012) Global Diversity of Sponges (Porifera). PLoS ONE 7(4): e35105. doi:10.1371/journal.pone.0035105
[43] Evans, J. Sea Remedies: Evolution of the Senses. 2009. Emryss. P153.
[44] Mangialavori, M. Koine International R-Z I. Spongia Toasta Entry. Accessed On Referenceworks.
[45] IBID.
[46] Mehbub et al. Marine Sponge Derived Natural Products between 2001 and 2010: Trends and Opportunities for Discovery of Bioactives. Mar. Drugs 2014, *12*, 4539-4577.
[47] IBID.
[48] Zeff, Jared. Practical Homeopathy. Online course. May 2016. http://www.medicinetalkpro.org/continuing-education/shop-online-ce
[49] Lemloh, M. et al. Diversity and abundance of photosynthetic sponges in temperate Western Australia. *MC Ecology* 2009, **9**:4
[50] Onthank, K.L. http://commons.wikimedia.org/wiki/File:Spongilla_lacustris.jpg
[51] Femmer, R. Online Image Accessed Feb 13th 2015. http://commons.wikimedia.org/wiki/File:92_ANM_Glass_sponge_2.jpg
[52] Anonymous. "Watering Sponge (Euplectella aspergillum)" . July 30 2013. Last Accessed October 19th 2014. Online Document accessed at https://translate.googleusercontent.com/translate_c?depth=1&hl=en&ie=UTF8&prev=_t&rurl=translate.google.com&sl=auto&tl=en&u=http://c3-in-

berlin.blogspot.de/2013/07/giekannenschwamm-euplectella-aspergillum.html&usg=ALkJrhgYOczUAv1xqusLuFB95bxjlZuCMw

[53] IBID

[54] IBID

[55] Image from Picchetti, G. http://commons.wikimedia.org/wiki/File:Spongia_officinalis.jpg

[56] Hahnemann, S. Hempel, J (trans). Materia Medica Pura. 1846. Accessed on Reference Works. Spongia Entry.

[57] Schlingensiepen-Brysch, I. The Source in Homeopathy: Cosmic Diversity and Individual Talent. 2009. Narayana Verlag. Kandern, Germany. Pp. 215-35.

[58] Lalor, L. "A Spongia Case- An Epic Journey out of Persecution from the Past" Homoeopathic Links 2013; 26(3): 177-180.

[59] Lalor, L. "A Spongia Case- An Epic Journey out of Persecution from the Past" Homoeopathic Links 2013; 26(3): 177-180.

[60] Lalor, L. "A Spongia Case- An Epic Journey out of Persecution from the Past" Homoeopathic Links 2013; 26(3): 177-180.

[61] Schlingensiepen-Brysch, I. The Source in Homeopathy: Cosmic Diversity and Individual Talent. 2009. Narayana Verlag. Kandern, Germany. Pp. 215-35.

[62] Lalor, L. "A Spongia Case- An Epic Journey out of Persecution from the Past" Homoeopathic Links 2013; 26(3): 177-180.

[63] Pechenik, J. Biology of the Invertebrates (4th ed). 2000. McGraw Hill. Toronto. Pp 95.

[64] IBID Pp 95-8.

[65] IBID P 96.

[66] IBID P 96.

[67] IBID P 102.

[68] IBID.

[69] Tow basic body forms of cnidaria: medusa and polyp. Ref: *Ruppert, E.E., Fox, R.S., and Barnes, R.D. (2004) Invertebrate Zoology (7th ed.), Brooks / Cole, pp. 111-124* http://upload.wikimedia.org/wikipedia/commons/1/17/Cnidaria_medusa_n_polyp.png

[70] Schleiden M. J. "Die Entwicklung der Meduse". In: "Das Meer". Verlag und Druck A. Sacco Nachf., Berlin, 1869. http://upload.wikimedia.org/wikipedia/commons/6/61/Schleiden-meduse-2.jpg

[71] Pechenik, J. Biology of the Invertebrates (4th ed). 2000. McGraw Hill. Toronto. Pp 116.

[72] Pechenik, J. Biology of the Invertebrates (4th ed). 2000. McGraw Hill. Toronto. Pp 110-2.

[73] IBID.

[74] Fred Hsu. Online Image Accessed June 1st 2017. https://commons.wikimedia.org/wiki/File:Orange_Sea_Pen_Monterey_Bay_Aquarium.jpg

[75] Pechenik, J. Biology of the Invertebrates (4th ed). 2000. McGraw Hill. Toronto. Pp 103.

[76] Blakely, Sierra. Online image accessed February 20th 2015. http://en.wikipedia.org/wiki/File:Aequorea3.jpeg

[77] Pechenik, J. Biology of the Invertebrates (4th ed). 2000. McGraw Hill. Toronto. Pp 98.

[78] IBID Pp 98-102.

[79] Acharya, S. Online Image accessed February 20th 2015. http://en.wikipedia.org/wiki/File:Chrysaora_Colorata.jpg

[80] Pechenik, J. Biology of the Invertebrates (4th ed). 2000. McGraw Hill. Toronto. Pp 102-3.

[81] IBID.

[82] Forgerz. Online image accessed February 20th 2015. http://en.wikipedia.org/wiki/File:Cubozoas.JPG

[83] Layne, M. Online image accessed February 20th 2015. http://en.wikipedia.org/wiki/File:Haliclystus_stejnegeri_1.jpg

[84] Pechenik, J. Biology of the Invertebrates (4th ed). 2000. McGraw Hill. Toronto P 110.

[85] IBID P 98-102.

[86] IBID P 102-3.

[87] Marques, A. & Collins, A. Cladistic analysis of Medusozoa and cnidarian evolution. Invertebrate Biology 123 (1). pp. 23–42.

[88] Anonymous. Cnidaria: The Fossil Record. Online document accessed Dec 15th 2014. Available at: http://www.ucmp.berkeley.edu/cnidaria/cnidariafr.html

[89] Borchiellini, C., Manuel, M., Alivon, E., Boury-Esnault, N., Vacelet J., and Le Parco, Y. (2001). "Sponge paraphyly and the origin of Metazoa". Journal of Evolutionary Biology 14 (1): 171–179.

[90] Hsieh, P. Fui-Ming, L. Rudloe, J. Jellyfish as Food. May 2001, Volume 451, Issue 1-3. Pp 11-7.

[91] IBID.

[92] Evans, J. Sea Remedies: Evolution of the Senses. 2009. Emryss. Pp 177-9.

[93] IBID.

[94] IBID.

[95] IBID.

[96] Allen, TF. Allen's Encyclopedia. Medusa and Corallium rubrum Entries. Accessed on ReferenceWorks.

[97] Chauhan, D. "I Have to Make an Effort to Maintain my Structure, Otherwise I Melt" Homeopathic Links. Spring 2014. Vol 27: 21-5.

[98] Evans, J. Sea Remedies: Evolution of the Senses. 2009. Emryss. Pp 161-2.

[99] IBID P 162.

[100] Thank you to the following website for these suggestions:
Wojcik, J. The Coolest and Strangest Jellies. Online Document last accessed Dec 17th 2014. http://bogleech.com/bio-jelly2.html

[101] Shebs, S. http://commons.wikimedia.org/wiki/File:Anthopleura_xanthogrammica_1.jpg

[102] Shepard, C. Homeopathic Proving of Anthopleura xanthogrammica. 2005. Online Document Accessed Dec 18th 2014.
http://www.homeopathycourses.com/pdf/GiantGreenSeaAnemone.pdf
[103] Evans, J. Sea Remedies: Evolution of the Senses. 2009. Emryss. Pp 191-209.
[104] Geunther, R. The Giant Green Sea Anenome. Online Document last accessed Dec 18th 2014. Available at http://www.victoriahomeopathy.com/sea-anemone.html
[105] Shepard, C. Homeopathic Proving of Anthopleura xanthogrammica. 2005. Online Document Accessed Dec 18th 2014. Pp 34-5.
http://www.homeopathycourses.com/pdf/GiantGreenSeaAnemone.pdf
[106] Taken from Evans, J. Sea Remedies: Evolution of the Senses. 2009. Emryss. Pp 191, 201.
[107] Image of Antipathes dendrochristos. Amend, M.
http://commons.wikimedia.org/wiki/File:Antipathes_dendrochristos.jpg
[108] Wichman, J. Corallium nigrum entry on Provings.info. Online Document Accessed Jan 2nd 2015.
http://provings.info/en/substanz.html?substanz=Corallium%A0nigrum&proving=hide
[109] Image by Zell. H.
http://commons.wikimedia.org/wiki/File:Corallium_rubrum_01.JPG
[110] Evans, J. Sea Remedies: Evolution of the Senses. 2009. Emryss. Pp 177.
[111] IBID Pp 169-70.
[112] Murphy, R. Nature's Materia Medica. 2007. Corallium rubrum entry. Accessed On Reference Works.
[113] IBID.
[114] Evans, J. Sea Remedies: Evolution of the Senses. 2009. Emryss. Pp 170.
[115] Asman, P & Lenoble, J.
http://commons.wikimedia.org/wiki/File:Spotted_trunkfish_Lactophrys_bicaudalis_%282412821171%29.jpg
[116] Schadde, A. Diploria Clivosa – Hirnkoralle. Online document last accessed Dec 18th 2014. Available at: http://www.anne.schadde.de/cms/hirnkoralle.html
[117] Schadde, A. Diploria Clivosa – Hirnkoralle. Online document last accessed Dec 18th 2014. Available at: http://www.anne.schadde.de/cms/hirnkoralle.html
Translated by Google Translate. Edited for grammar by Dr. Paul Theriault. With thanks to Vera Jamin for translating some of the more difficult sentences.
[118] Anonymous. Dimorphaestra d'Orbigny 1850 (stony coral). Online Document accessed May 31st 2015. http://fossilworks.org/?a=taxonInfo&taxon_no=6358
[119] Mercy, M. Homeopathic Remedies from the Fossil Kingdom. 2011. L'Ile Au Phare.
[120] Mercy, M. Remedies from the Past for Toxicity Today. 2017. L'Ile Au Phare.Pp 108-119.
[121] Zander, J. Online Image Accessed June 2nd 2015.
http://en.wikipedia.org/wiki/File:Mushroom_Coral_%28Fungia%29_Top_Macro_91.JPG
[122] Mercy, M. Fascinating Fossils: New Homeopathic Remedies. 2009. L'Ile Au Phare. Pp. 213-226.

[123] Mercy, M. Homeopathic Remedies from the Fossil Kingdom. 2011. L'lle Au Phare.
[124] Image from Haplochromis. http://commons.wikimedia.org/wiki/File:Heteractis_malu.JPG
[125] Begin, M. Heteractis malu; The Sea Anemone: Exposed and vulnerable, must protect by withdrawal inward. InterHomeopathy. March 2009. Online Document Last accessed Dec 18th 2014. Available at http://www.interhomeopathy.org/sea_anemone_must_protect_by_withdrawal_inwardly
[126] Image from Arvelund, M. http://commons.wikimedia.org/wiki/File:A_ocellaris_2_Sesoko_Point_140904_LOWRES.jpg
[127] Burch, M & Sieben, S. Stoichactis Kenti Sea Anenome Proving. 2006. Online Document Accessed Dec 19th 2014. Available at http://provings.info/pruefungen/stichodactyla-sieben-en.pdf
[128] Fautin, G. &Allen, G. Anemone Fishes and Their Host Sea Anemones (2 (1997) ed.). Western Australian Museum. p. 160.
[129] Burch, M & Sieben, S. Stoichactis Kenti Sea Anenome Proving. 2006. Online Document Accessed Dec 19th 2014. Available at http://provings.info/pruefungen/stichodactyla-sieben-en.pdf
[130] Evans, J. Sea Remedies: Evolution of the Senses. 2009. Emryss. Pp 211- 219.
[131] Burch, M & Sieben, S. Stoichactis Kenti Sea Anenome Proving. 2006. Online Document Accessed Dec 19th 2014. Available at http://provings.info/pruefungen/stichodactyla-sieben-en.pdf . P 21.
[132] Image from Gautsch, G. http://commons.wikimedia.org/wiki/File:Avispa_marina_cropped.png
[133] Horton, J. Do Jellyfish have the Deadliest Venom in the World? Online document last accessed Jan 2 2014. http://animals.howstuffworks.com/marine-life/jellyfish-venom2.htm
[134] Grey, A. The Experience of Medicine Vol 1. 2005. Emyryss. Pp. 200-69.
[135] Hillewaert, H. http://commons.wikimedia.org/wiki/File:Aurelia_aurita_2.jpg
[136] Shoji, J.; Yamashita, R.; Tanaka, M. Effect of low dissolved oxygen concentrations on behavior and predation rates on fish larvae by moon jellyfish Aurelia aurita and by a juvenile piscivore, Spanish mackerel Scomberomorus niphonius. 2005. Marine Biology 147 (4). pp. 863–868.
[137] Evans, J. Sea Remedies: Evolution of the Senses. 2009. Emryss. P 232.
[138] IBID Pp 231-41.
[139] I have also used data from Murphy R. Narure's Materia Medica. 2007. Medusa Entry. Accessed Via Reference Works.
[140] Chauhan, D. I Have to Make an Effort to Maintain my Structure, Otherwise I Melt. Homeopathic Links 2014 27(1): 21-25.
[141] Image from Yuksel, V. http://commons.wikimedia.org/wiki/File:Portuguese_Man_o%27_War_at_Palm_Beach_FL_by_Volkan_Yuksel_DSC05878.jpg
[142] Evans, J. Sea Remedies: Evolution of the Senses. 2009. Emryss. P 227.

[143] IBID P 221.
[144] IBID.
[145] Adam, G. I'm on the ferry with lots of my family and we lose each other: A Case of Physalia Physalis. Interhomeopathy. March 2013.
http://www.interhomeopathy.org/im-on-the-ferry-with-lots-of-my-family-we-lose-each-other-a-case-of-physalia-physalis
[146] Pechenik, J. Biology of the Invertebrates (4th ed). 2000. McGraw Hill. Toronto. Pp 127.
[147] IBID. P 132.
[148] Raskoff, K. Online Image accessed March 26th 2016.
https://en.wikipedia.org/wiki/File:LightRefractsOf_comb-rows_of_ctenophore_Mertensia_ovum.jpg
[149] IBID. Pp 128.
[150] Franc, J. Organization and Function of Ctenophore Colloblasts: An ultrastructural Study. Biol Bulletin. Vol 155 No 3 (Dec 1978) Pp 527-44.
[151] Griswold, R. Online Image Accessed March 27th 2016.
https://en.wikipedia.org/wiki/File:Ctenophore.jpg
[152] Schnitzler, C et al. Genomic organization, evolution and expression of photoprotein and opsin genes in Mnemiopsis leidyr: a new vier of Cnetophore photocytes. BMC Biology 2012. 10: 107.
[153] Pechenik, J. Biology of the Invertebrates (4th ed). 2000. McGraw Hill. Toronto. Pp 132.
[154] IBID Pp 128.
[155] Philcha. Online Image Accessed March 27th 2016.
https://en.wikipedia.org/wiki/File:Ctenophore_body_vert_section.png
[156] Hobgood, N. Online Image accessed March 27th 2016.
https://en.wikipedia.org/wiki/File:Coeloplana_astericola_%28Benthic_ctenophores%29_on_Echniaster_luzonicus_%28Seastar%29.jpg
[157] Anderson, A. Online Image Accessed March 27 2016.
https://en.wikipedia.org/wiki/File:Zooplankton2_300.jpg
[158] Qu et al. A Vanished history of skeletonization in Cambrian comb jellies. Scientific Advances 10 Jul 2015. Vol 1 no 6.
http://advances.sciencemag.org/content/1/6/e1500092.full
[159] Apokryltaros. Online Image Accessed March 27th 2016.
https://en.wikipedia.org/wiki/File:Maotianoascus_octanarius.JPG
[160] Haddock, S. Comparative feeding behavior of Planktonic Ctenophores. Integrative and Comparative Biology Vol 47 issue 6 Pp 847-53.
[161] Qu et al. A Vanished history of skeletonization in Cambrian comb jellies. Scientific Advances 10 Jul 2015. Vol 1 no 6.
[162] Morris, S. Collins, D. Middle Cambrian Ctenophores from the Stephen Formation, British Columbia, Canada. Philosphical Transactions of the Royal Society of Biological Sciences. March 1996 Vol 351 Iss 1337.
[163] Ghedoghedo. Online image Accessed March 27th 2016.
https://en.wikipedia.org/wiki/File:Eoandromeda.jpeg

[164] Tang et al. Eoandromeda and the Origin of the Ctenophora. Evolution and Development. Vol 13 Issue 5.
[165] Shu et al. Lower Cambrian Vendobionts from China and Early Diploblast Evolution. Science 05 May 2006 Vol 312 Issue 5774 Pp 731-734.
[166] IBID.
[167] Apokryltaros. Online Image Accessed March 27th 2016.
https://en.wikipedia.org/wiki/File:Stromatoveris_psygmoglena.jpg
[168] Moroz et al. The ctenophore genome and the evolutionary origin of nervous systems. Nature 510 109-14. June 5th 2014.
[169] Reitzel et al. Nuclear receptors from the ctenophore *Mnemiopsis leidyi* lack a zinc-finger DNA-binding domain: lineage-specific loss or ancestral condition in the emergence of the nuclear receptor superfamily? EvoDevo. 2011; 2: 3.
[170] Maxwell et al. MicroRNA and essential components of the microRNA processing machinery are not encoded in the genome of the ctenophore Mnemiopsis leidyi. BMC Genomics. 2012; 13: 714.
https://www.ncbi.nlm.nih.gov/pmc/articles/PMC3563456/
[171] Pang et al. Genomic insights into Wnt signaling in an early diverging metazoan, the ctenophore *Mnemiopsis leidyu*. EvoDevo. 2010; 1: 10.
[172] Kohn et al. Rapid evolution of the compact and unusual mitochondrial genome in the Ctenophore Pleurobranchia bachei. Molecular Phylogenetics and Evolution. Vol 63, Iss 1. April 2012. Pp 203-207.
[173] Moroz et al. Nature 509,411. (22 May 2014)
[174] Whelan, N et al. Error, signal and the placement of the Ctenophora sister to all other animals. Proceedings of the National Academy of Sciences of the United States of America. Vol 112 no 18 Pp 5773-8.
[175] Borowiec, M. Extracting phylogenetic signal and accounting for bias in whole genome data sets supports the Ctenophora as sister to the remaining metazoa. BMC Genomics. 2015. 16: 987.
http://bmcgenomics.biomedcentral.com/articles/10.1186/s12864-015-2146-4
[176] Anonymous. Online Image Accessed March 27th 2016.
https://commons.wikimedia.org/wiki/File:Warty_comb_jelly_%28mnemiopsis%29.jpg

Appendix: Charts
Sponges:

| Porifera | Existing or not existing based on environmental factors. Does the being have the things necessary to incarnate into the world? Deep insecurity about existing without specific things. Affinity to the glandular system. |

Sponge class	Possible themes
Hexactinillids (Glass Sponges)	The formation of syncytia (large multinucleated cells formed from the merging of smaller cells with single nuclei) suggests to me that the process of incarnation may be resisted in this group. They may wish to return and remerge with the divine. I suspect each individual remedy will have a particular issue that causes them the retreat back into the oneness out of which they emerge.
Desmosponges	The being is unsure of whether or not to commit to life. They perceive some of the necessary factors for being alive to be deficient, and thus avoid making a commitment and fully incarnating. Each individual remedy likely has a separate issue that stops them from fully committing to life.
Calcareous	The being feels ready to commit to life, but issues of inadequacy in some respect keep them from doing so. Each individual remedy likely has a separate inadequacy that prevents them from fully incarnating.

Sponge class	Proven Remedies	Unproven Remedies
Glass (Hexactinellid) Sponges	Euplectella aspegillium (euple-a)[V]	None
Desmosponges	Spongia Toasta (spong), Badiaga (bad)	Aplysina aerophoba (aply-a)[R], Chondosia reniformis (chond-r)[R], Tectitethya crypta (tect-c)[A]
Calcareous Sponges	None	Clathrina clathrus (clath-c)[R]
A- indicates the Remedy is available from Ainsworth's R- indicates the remedy is available from Remedia V- indicates the remedy is available from Verfügbarkeit des Mittels bei Enzian Apotheke erfragen All other remedies are available at most pharmacies.		

Summary of Key Issues of Sponges:

Remedy	Main Issue within the Table
Badiaga	I cannot incarnate because the world is a terrible place, full of danger and insecurity. I lack the security I need to incarnate.
Euplectella aspergilum	I do not wish to incarnate because I cannot deal with the mental effort required in life.
Spongia Toasta	I lack the social support I need to exist.

Ctenophora:

| Ctenophora | Existing or not existing based on sensation. Exerting effort on one's own behalf. Wishes to simply experience life without any effort or activity, but does not wish to have unpleasant experiences and so must exert effort to avoid them. Exerting this effort exhausts the being. |

Proved Remedies	Unproved Remedies
Mnemiopsis macrydi[IE]	None
IE- Available from I & E Organics	

Cnidarians:

Cnidaria	Have not built the boundary between themselves and the outside world. Cannot separate themselves from the outer world, and exposure to the outer world dissolves their sense of self.

Subphyla	Themes
Anthozoa	Experience the overload of sensory input from the outside world and withdraw from it into a quiet and less intense environment.
Medusozoa	Experience overload of sensory input from the outside world and must cope with it and adapt to it rather than withdrawing.

Subphyla	Proved Remedies	Unproved remedies
Anthozoa	Anthopleura xanthogrammaica[G] (antho-p), Corallium rubrum[G] (cor-r, Gorgonio nobilis), Corallium nigrum[H] (cor-n), Diploria clivosa[E] (diplo-c), Fossil Fungia Coral[A], Fossil Coral Dimorphophaestra[A], Heteractis malu[R,F] (hetera-ma), Stichodactyla gigantea (stich-g)[H,F]	Calliactis parasitica[R] (callis-p), Eunicella singularis[R] (euni-s)
Hydrozoa	Physalia physalis[R,F,H] (physala-p)	Hydra oligactis[R] (hydra-o)
Scyphozoa	Medusa[G] (medus, Aurelia aurita)	None
Cubozoa	Chironex fleckerii[H,F,R] (chir-f)	None
Staurozoa	None	None
R - Available at Remedia F – Available at Freemans G - Available at most pharmacies H - Available at Helios E – Available at Enzian		

Summary of the Key Issues Of Cnidarians:

Remedy	Main Issue Within The Table
Anthozoa	
Anthropleura xanthogrammatica	The being feels profound connection with others which stabilizes their sense of self. Anything that disrupts their connection to others, especially taboo and unacceptable behaviors, or social conflict, disrupts their connection and their sense of self.
Corallium nigrum	Social conflict and having others make social demands which tax the beings inner reserves, disrupts the being's sense of self. The response is to withdraw from conflict and isolate oneself.
Corallium rubrum	The being feels their sense of self dissolving because of the intensity of sensory stimulation the receive from the external world. They will respond by hiding and withdrawing.
Diploria clivosa	The being requires symbiosis with lifeforms around it in order to maintain the integrity of its self. This symbiosis is disrupted by chaos, brought on by willfulness, going against the higher divine will.
Fossil Dimorphastrea	The being feels extremely intoxicated by its environment, both environmentally and socially. It withdraws into seclusion and isolation. It perceives this isolation as lonely however, wishing to have lively exchange with the outside world.
Fossil Fungia coral	The being is immersed in its ancestral inheritance so completely that it is unable to completely live its own life. It floats above all involvement with its life, and is unable to fulfill the roles of its current life, or even truly feel it.
Heteractis malu	Conflict disrupts the beings sense of self, so it withdraws and gives in to the demands of others to avoid it.
Stichodactyla gigantea	The being requires support from it's social circle to maintain their sense of selves. It requires others to adhere to their moral codes and have their sense of self is disrupted when others aren't acting according to the being's ideals. The being usually perceives this as acting against their own personal integrity. They will either become quite irritable or withdraw in response to this.
Medusozoa	
Chironex fleckerii	Disorder disrupts the integrity of the beings sense of self, so the being actively works to create order in its life. Aggression is often resorted to in order to restore order. Dissolution occurs when the being is exhausted an unable to order its world, or it is surrounded by to much disorder.
Medusa	Change disrupts the beings sense of self-integrity, and so they actively create a stable environment for itself. If change exceeds the beings ability to cope, it goes into crisis and feels itself dissolve.
Physalia physalis	The being feels that it needs to maintain an understanding of the world around it. Confusion and feeling stupid destroy its sense of self.

Layer	1	2	3	4	5	6	7
Parazoa	Porifera						
Ctenophora	Ctenophora						
Cnidaria		Cnidaria					
Lophoptrochozoa		Brachiopods	Annelida	Mollusca	Platyhelminthes		
			Bryozoa				
Ecdysozoa			Nematoda		Arthropoda	Arthropods-Endopterygotes	Arthropods-Mecopterygotes
					Trilobites	(Beetles, Bees)	Amphismenoptera
					Chelicerates		(Butterflies, Flies)
					Crustaceans		
					Early Insects		
Deuterosomes	Lancelets	Echinoderms	Amphibians	Dinosaur		Mammals	Birds
	Tunicates	Ray Fish	Turtles				
	Hemichordata	Cartilage Fish	Squamata				
			Crocodilians				
Human remedies	Lac humanum	Vernix caseosa	Meconium humanum	Placenta	Aqua amniota	Umbilicus	
	Lac maternum		CCC				
			Carcinosinum				

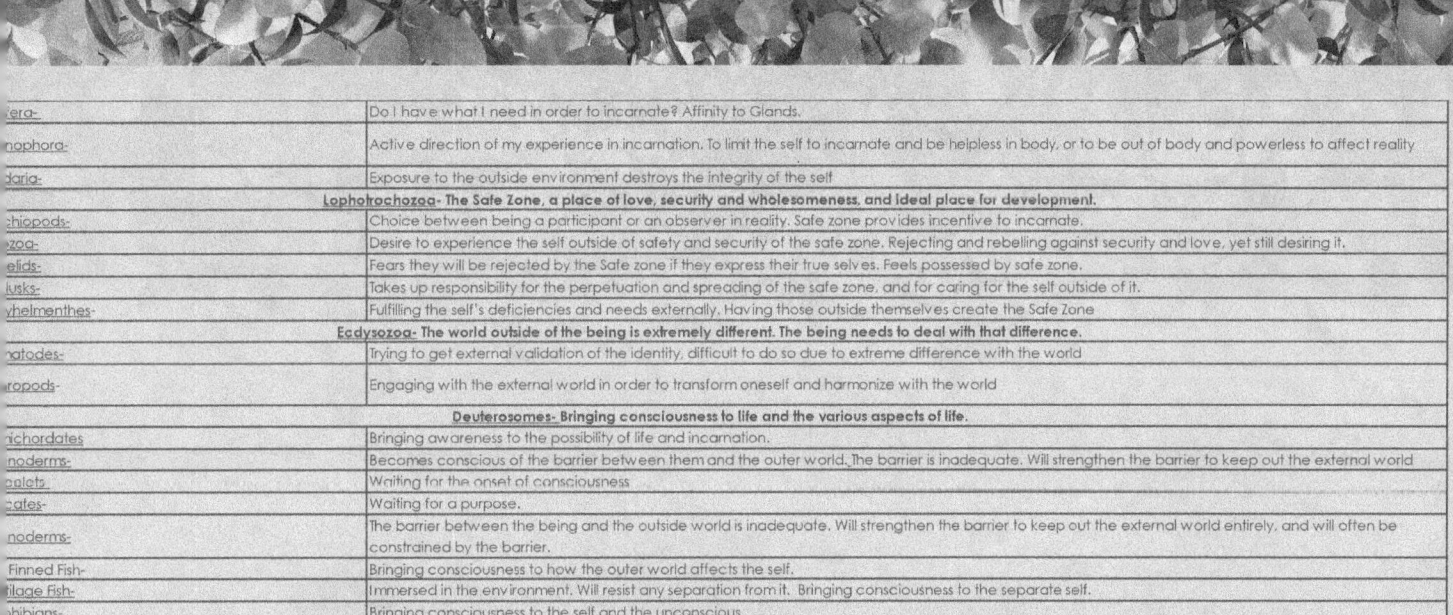

...fera-	Do I have what I need in order to incarnate? Affinity to Glands.
...nophora-	Active direction of my experience in incarnation. To limit the self to incarnate and be helpless in body, or to be out of body and powerless to affect reality
...daria-	Exposure to the outside environment destroys the integrity of the self
Lophotrochozoa- The Safe Zone, a place of love, security and wholesomeness, and ideal place for development.	
...chiopods-	Choice between being a participant or an observer in reality. Safe zone provides incentive to incarnate.
...zoa-	Desire to experience the self outside of safety and security of the safe zone. Rejecting and rebelling against security and love, yet still desiring it.
...elids-	Fears they will be rejected by the Safe zone if they express their true selves. Feels possessed by safe zone.
...lusks-	Takes up responsibility for the perpetuation and spreading of the safe zone, and for caring for the self outside of it.
...yhelmenthes-	Fulfilling the self's deficiencies and needs externally. Having those outside themselves create the Safe Zone
Ecdysozoa- The world outside of the being is extremely different. The being needs to deal with that difference.	
...natodes-	Trying to get external validation of the identity, difficult to do so due to extreme difference with the world
...ropods-	Engaging with the external world in order to transform oneself and harmonize with the world
Deuterosomes- Bringing consciousness to life and the various aspects of life.	
...richordates	Bringing awareness to the possibility of life and incarnation.
...noderms-	Becomes conscious of the barrier between them and the outer world. The barrier is inadequate. Will strengthen the barrier to keep out the external world
...plets	Waiting for the onset of consciousness
...ates-	Waiting for a purpose.
...noderms-	The barrier between the being and the outside world is inadequate. Will strengthen the barrier to keep out the external world entirely, and will often be constrained by the barrier.
...Finned Fish-	Bringing consciousness to how the outer world affects the self.
...ilage Fish-	Immersed in the environment. Will resist any separation from it. Bringing consciousness to the separate self.
...phibians-	Bringing consciousness to the self and the unconscious
...es-	The outside world is harmful to the self. The being armors itself from the outside world, but feels trapped.
...amata	The outside world is harmful to the self. The being uses subterfuge to diffuse its impact on the itself.
...codilians	The outside world is harmful to the self. The being deals with this by maintaining control over the outside world.
...saurs	Bringing consciousness to ones social environment.
...mmals	The self versus the group. Bringing consciousness to one's interaction's with the group and the nature of the group.
...s	Bringing consciousness with what separates the self from freedom and perfection

Naturopathic Doctor

www.ingramcontent.com/pod-product-compliance
Lightning Source LLC
Chambersburg PA
CBHW080923170526
45158CB00008B/2212